DC ARC ANALYSIS

PHILIPS TECHNICAL LIBRARY

DC ARC ANALYSIS

DR. N. W. H. ADDINK

Formerly Head of the Spectrochemical Department
Philips Research Laboratories
Eindhoven, The Netherlands

MACMILLAN EDUCATION

Original Dutch edition © N. V. Philips' Gloeilampenfabrieken, Eindhoven
English edition © N. V. Philips' Gloeilampenfabrieken, Eindhoven, 1971
Softcover reprint of the hardcover 1st edition 1971

333 13114 2

ISBN 978-1-349-15415-9   ISBN 978-1-349-15413-5 (eBook)
DOI 10.1007/978-1-349-15413-5
First published in English by
THE MACMILLAN PRESS LTD
London and Basingstoke
Associated companies in New York, Toronto, Melbourne
Dublin, Johannesburg and Madras

 **PHILIPS**

Trademarks of N. V. Philips' Gloeilampenfabrieken

Dedicated to the Institution
Colloquium Spectroscopicum Internationale

Contents

Preface

After we had found, in 1953, that the K- and Q-methods of analysis using a spectrograph could give quantitative results for a wide variety of inorganic materials, burnt in a d.c. carbon arc, introductory papers about the method were published in German and English to summarize the work so far [a, b, c]. Now a more detailed account of the method is in this book, written, as may be understood, for spectrochemists.

There are many ways of tackling the reporting of a problem, the usual one of beginning by describing all its complexities, and the reasons why the results can never be quite certain, is unsuited to my nature; and I would prefer to quote the following passage from Goethe's Faust, Part I, where Faust in his study says:

Geschrieben steht: "Im Anfang war das Wort!"	(line 1224)
Hier stock' ich schon! Wer hilft mir weiter fort?	(line 1225)
. . . . . .	
Mir hilft der Geist! Auf einmal seh' ich Rat	(line 1236)
Und schreibe getrost: Im Anfag war die Tat!	(line 1237)

So there are precedents for action coming before theory, although this way of working has its drawbacks. If the work is successful, the theoretical justification of empirically derived relationships is liable to seem redundant; for example, originally the relationship between evaporation rate and boiling point was only schematically outlined, and more precise derivation has been added during working up notes into this book. In demand for speed characteristic of industrial laboratory work, one has to produce results and some estimate of their reliability; and I have therefore tried to publish general methods of working for d.c. arc analysis, and also for X-ray fluorescence [d], mass spectrometry [e], flame photometry [f] although in the last case the publication is only a preliminary one.

I wonder how the K- and Q-method will work out in other laboratories? I expect criticisms that the exposure times are too long, that the book is not a 'cookery book', and that the method needs some skill; but I have tried to show that after reading the description given

here the mystery disappears. Instead of speculating further I prefer
to quote the words of Ahrens [g] at the beginning of his book:

'and so, back to work!'

Oosterbeek (Gld.)   N. W. H. Addink
The Netherlands 1969

References

a. Addink, N. W. H., *C.S.I. III High Leigh* 1952, *Spectrochim,
   Acta* (1953), pp. 495–9.
b. Addink, N. W. H., Dikhoff, J. A. M., Frl. Schipper, C., Witmer,
   A. W. and Groot, T., *Spectrochim, Acta* (1955), **7**, pp. 45–59.
c. The same, *Applied Spectroscopy*, 1956, **10**, 128–37.
d. Witmer, A. W. and Addink, N. W. H., *Science and Industry*,
   12, 1–6 (1965).
e. Addink, N. W. H., *Z.f. Anal Chemie*, **206**, 81–8 (1964).
f. Addink, N. W. H., *C.S.I. XV, Madrid*, 1969, Abstract No. 251;
   Proceeding II, pp. 47–54.
g. Ahrens, L. H., *Spectrochemical Analysis*, Addison–Wesley
   Press 1950, p. xix.

Acknowledgements

I am greatly indebted to the Directors of Philips Research Laboratories, Eindhoven, Netherlands, Dr. E. J. W. Verwey and Dr. P. W. Haayman, for the great latitude they allowed me in developing analysis methods and carrying out experiments leading to results described in this book.

Prof. Dr. J. H. de Boer (great 'amphilysator' as he called himself) has said in his valedictory address: 'Applied science cannot exist without pure Science'. In connection with this remark I should like to mention the names of Dr. W. de Groot and Dr. F. M. Penning,† always willing to discuss problems in relation to pure science.

In another sense reading of theses originating from the School of Ornstein has given me much information.

It is understandable that during the years 1948–67 many co-workers have given their help in carrying out research work, new analysis work and routine-analysis work. Their names are mentioned here:

Miss E. Aarts 1963–4
Miss M. Baten 1958–
Miss B. van Batenburg 1952–5
Mr. van Beek 1958–60
Mr. B. van den Berg 1952–
Miss J. Bol. 1951–3
Miss J. de Bont 1966–7
Miss M. Briene 1949–53
Mr. A. van den Broecke 1947–8
Mr. O. Büdgen 1959–
Mr. J. A. M. Dikhoff 1951–8
Mr. J. van Dorenmalen 1948–51
Miss T. van den Eynden 1963–
Miss C. Groot 1954–8
Mr. T. Groot 1952–
Miss T. Heiligers 1959–65
Miss H. Hovingh 1955–9
Mr. J. van Kollenburg 1958–67

Mrs. M. Koopmans 1964–
Mrs. A. van den Kroonenberg 1960–1
Miss T. Luykx 1953–5
Mr. L. Monten 1949–52
Mr. D. de Mooy 1965–
Miss H. Savenije 1955–66
Miss C. Schipper 1951–6
Mr. J. Sneyders 1957–
Miss A. Stapert 1949–51
Miss G. Stienstra 1953–64
Mr. B. Verkerk 1950
Mr. P. Verschueren 1954–
Mrs. C. Verspui 1960–4
Mr. A. W. Witmer 1950–
Miss Th. Zegeling 1955–62
Miss S. Zwijsen 1958–68

To all of them my sincere thanks for their helpful co-operation.

A spectrochemical laboratory cannot exist without help and direction from other laboratories, particularly the so-called wet-chemical lab. My gratitude to workers in the laboratory of Dr. A. Claassen and his co-workers Messrs. L. Bastings, F. de Boer, J. Knappe and J. Visser.

Prof. Dr. H. J. Eichhoff (Gutenberg University, Mainz, W. Germany) was kind enough to read through the manuscript and I am grateful to him for his painstaking review. I remember with pleasure the time spent in his laboratory where we discussed many problems (a result of this co-operative work can be found in the Proceedings of C.S.I. VIII, Lucerne 1959, Verlag Sauerländer, Aarau (Switzerland) 1960, pp. 89–92). I am also grateful to him for his idea to have data cards for each element. They will be easy to use.

Dr. H. Nickel (Institut für Reaktorwerkstoffe, Kernforschungs-anlage Jülich, W. Germany) has read through Section 1.3 and I have had the tremendous benefit of his work on chemical reactions with graphite. We are both sure that values of melting points and boiling points of the elements, their oxides and their carbides have to be considered with a pinch of salt! But one can also be grateful to those who show us the significance of the uncertainty.

To Mr. E. H. S. van Someren, B.Sc., F.Inst.P., well-known to researchers in spectrochemistry, my sincere thanks for the advice concerning wording and grouping of chapters.

And last but not least to Dr. S. D. Boon, Deputy Director of Centrex Publishing Company, and to his co-workers, my gratitude for their pleasant co-operation.

I had the opportunity of reading various papers on arc analysis, also on other subjects dealing with methods of analysing inorganic substances at meetings of the C.S.I. and once again I will thank participants of our colloquia for their interest in my lectures. This is the reason why I have dedicated this book to the Institution Colloquium Spectroscopicum Internationale.

N.W.H.A.

Introduction

Spectrochemical analysis methods by means of an electric arc are not non-destructive. A certain quantity of the material to be analysed is necessary. A certain quantity (5 or 10 mg) is evaporated in the arc and is therefore lost as a vapour into the surrounding air.

The advantage of the method described in the following sections is that it is not limited by the chemical state of the sample. Metals, oxides, salts, can all be analysed, from the high temperature of the burnt spot ('Brennfleck') on the anode. Removal of water of crystallization, for instance by heating to 120 °C until constant weight, is necessary before loading the anode. Care must be taken with some metal halides and other volatile substances as they can disturb a smooth vaporization.

It is understandable that because a *general method* was aimed at, a compromise had to be reached. As boiling points of metals may vary from about 300 °C (b.p. of P) up to about 6000 °C (b.p. of W), exposure times diverge from 1 minute to 8 minutes maximum. So to fulfill the basic principle chosen, a *complete evaporation* either of main components and impurities, or of impurities only is required, if the main component is of a high-boiling refractory nature. Preburn is not put into practice.

There is a great difference between introducing samples into the radiating volume of the arc and into the flame in flame photometry. In the latter case the sample is in solution and sprayed through the flame, at the same initial speed for all constituents. The initial speed of evaporation of a small sample through the arc depends upon the boiling points of its constituents: a volatile substance will show a high speed of evaporation and a refractory substance a low speed one.

In order to obtain a smooth volatilization *large electrodes* (diameter 8 or 10 mm, length 35 mm) have been chosen as they prevent an explosive vaporization. A smooth vaporization is a consequence of a relatively slow increase of the temperature of anode tip and analysis material during arcing. Most substances fulfill these requirements, but some low boiling substances still evaporate in a less controllable way and high boiling materials do not evaporate completely. If the anode tip is smaller (diameter 4 mm) completion of evaporation has been achieved as can be seen from the following data (see also [27]):

Boiling point of various samples	Evaporation of a 5 mg sample from large and small anodes	
	diameter 10 mm	diameter 4 mm
about 3000 °C	100%	100%
about 4500 °C	40%	100%
about 5500 °C	13%	100%

All these facts, as well as reactions which take place in the crater of the anode of sample, electrode material (graphite) and air ('the carbon arc burns in air'!) have been taken into account in the development of the method. Recently, relevant work concerning reactions with carbon has been published [1], [2] and [40].

Two types of light source are used. In both cases, the arc current amounts to 10 ± 2 Amps during the whole arcing time.

In the first method (*K-method*) 5 mg of the sample are evaporated to completion through an arc gap of 9 mm. Only the light originating from the centre of the arc is let through to the spectrograph. In the second method (*Q-method*), 10 mg of the sample are evaporated through an arc gap of 2 mm and all the light (including that from cathode and anode layers) is presented to the optical instrument.

The core of the *K*-arc shows a mean gas temperature of 6100 K [3]; for the *Q*-arc this temperature amounts to about 6700 K [4]. In both cases the final temperature of the anode tip of 10 mm diameter is about 3500 °C and about 4000 °C or higher in case of 4 mm diameter.

If we now add to the problems mentioned before matrix effects, resulting in a change in the temperature of excitation, in a shifting of atom-ion equilibria or in carrier effects, we may frankly say: this is not a cookery book. We should consider the problems discussed in the following sections, otherwise nobody can dissuade the reader from carrying out the first experiment; let us say the analysis of a simple steel sample by applying the *K*-factors supplied in the Appendix. The result (see Section 3.3) will be of a semi-quantitative nature.

By looking through the comments in Section 4.2.2, one is automatically confronted with matrix effects just mentioned (Sections 2.4, 2.5 and 2.6) and with self-absorption (Section 2.3). Finally one may be using a spectrograph other than the *Hilger large quartz* (E 492); therefore read Section 3.4. Above all, one is interested in the origin of spectral lines; are they emitted from the core or from the mantle of the arc (Section 1.6) and this is of importance in the analysis results: the core protected against temperature fluctuations by the mantle will give more reproducible results than the mantle.

According to the experience of the author the accuracy (twice the coefficient of variation, c.o.v.) of the K-method amounts to 2–10 % relatively and of the Q-method results remain of a semi-quantitative nature, in both cases provided that duplicate exposures have been taken and measured. Accuracy can be improved in particular in the Q-method by taking spectra in six-fold. A further decrease of the limit of detection is described in Section 3.5

All factors influencing spectra originating from the sample brought into the arc have been outlined roughly in this introduction. That this is true has been confirmed by the determination of transition probabilities from K-factors. See [5] and table 8.

Chapter 1

Arcs and Atomization

The running of an arc was first observed by striking a discharge between two horizontal carbon rods, and as this took on a bowed shape due to the rising current of hot air, the arc got its name, and kept it even when it became vertical or ran in another gas; we still speak of it as 'burning' although only slight oxidation may occur. The electrode connected to the positive pole, the anode, becomes hotter than the cathode if their diameters are equal, and for most analytical work samples are placed in a crater on the anode. The temperature profile across the arc shows a central core (usually 2–4 mm diameter) considerably hotter than the surrounding mantle, which is also hot enough to be luminous, and has about three times as great a volume. In the arc chemical reactions may occur, which may either continue or reverse changes in the stage of combination of the elements leaving the crater. These reactions are briefly discussed in Section 1.3 and in greater depth for certain of the elements separately. The rate of evaporation of the element being analysed depends on (i) whether it evaporates as an element or a compound, (ii) on the temperature difference between its boiling point and that of the arc crater, and (iii) on the speed of evaporation of other substances present with it in the crater. These effects are discussed in more detail in Section 1.4 and individually for those elements of special importance. But the emission of radiation does not take place only from heated atoms, as from a metal filament, most of it is radiation from excited atoms and from ions, and the part played by the core of the arc and the ionization potential of individual elements in regulating this process is described in Sections 1.5 and 1.6. The theory quoted here is considerably simplified, but references to original papers are provided which enable readers to obtain deeper insight into excitation processes in the arc.

1.1. 'Burning' of a direct current arc

The striking of the arc is done by a short-circuit. The *short-circuiting current* is set to a predetermined value (*K*-arc: 16 A; *Q*-arc: 14 A). The cathode is then raised (the anode containing the sample is underneath) until the distance between the electrodes is 8½–9 mm (*K*-arc) or 2 mm (*Q*-arc). The exposure time now begins, during which the distances are maintained and optically (or automatically) monitored. What kind of process has taken place during opening of the arc gap?

Immediately after opening the arc gap, a spark discharge between cathode and anode was initiated through the air. Electrons, originating from the cathode, are accelerated by the high electric field and on arriving at the anode they lose their energy as heat. This is why the anode tip reaches a higher temperature than the cathode from which electrons are drawn during the whole process of evaporation.

Once the exposure time has begun the arc wanders and mainly burns the edges of the anode in various places. The discharge takes place between both graphite electrodes (C–C discharge). Sometimes glowing graphite particles can be observed. The temperature of the anode tip increases as does the temperature of the sample placed in the crater. After a certain time (short in the case of volatile substances, for instance 10 seconds; a longer one (1–2 minutes) in case of refractory materials) the boiling point or a temperature near to it is reached when the vapour pressure of the sample is sufficient to carry the arc discharge: 'the arc burns on the analysis material'. Depending on the electric circuit applied and the nature of the evaporating element the voltage between the electrodes can fall from 70 to 50 volts.

Expressed in physical terms, in the beginning during the C–C discharge, the arc temperature was limited by the dissociation energy of gases present in the arc gap (CO, NO, $\propto 10\frac{1}{2}$ eV); afterwards other molecules and atoms originating from the sample entered the arc gap and they all have taken over the discharge provided that their ionization potential (V_i) shows values smaller than 11 eV. This is the reason why only elements with a $V_i < 11$ eV can be analysed in a graphite arc 'burning' in air. See Fig. 1.

Rare gases, halogens, hydrogen, nitrogen and oxygen cannot be analysed in this way.

At the beginning of the discharge (C–C), the voltage between both electrodes amounted to 60–70 V (arc current about 8 A). During evaporation of the sample, the resistance of the arc diminishes and the voltage appears to drop to 50 V (arc current about 10 A) or less (arc current about 12 A), dependent upon the magnitude of

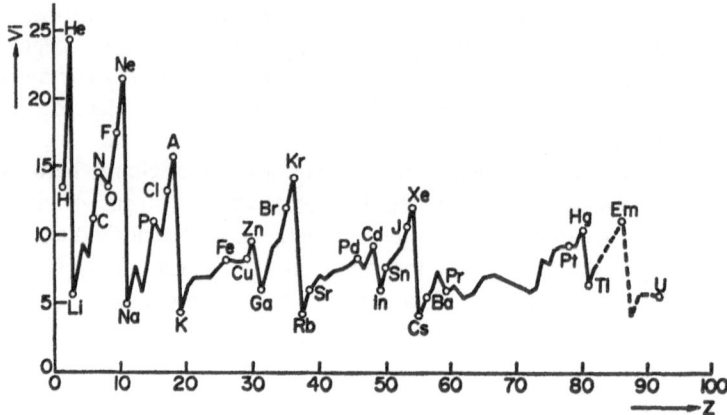

Fig. 1. The ionization energy V_i expressed in eV of the elements plotted as a function of their atomic number Z.

V_i of the evaporating element. The higher the electron concentration, the lower the resistance and also the voltage. During melting, boiling and evaporation of the sample it is likely that molten substances may penetrate into the anode and on this account, it is necessary to maintain a relatively long arcing time. At the end of that time, the resistance of the arc increases because of the pure C–C discharge and usually the arc current falls to about 8 A. In this period, the last vestiges of analysis material are vaporized. Variations in the current strength are attributable to this.

In order to keep the fluctuations at a minimum and to prevent the extinguishing of the arc at the end of the exposure time, it is recommended that *a driving voltage of some* 220 V be applied to the arc.

Spluttering of the sample to be analysed can take place mainly at two points of time: first, when the arc starts to strike on the analysis material and also in case of refractory materials when all impurities have been vaporized and the pure main component remains. It is important to observe the burning of the arc continuously through a dark glass in order to know what has happened (see further, Sections 2.6 and 4.1.5).

In low-pressure discharges, exchange of energy between electrons and gas atoms is small: 'electron temperature' is high, 'gas temperature' is low. However, at atmospheric pressure, this exchange of energy is so large that the gas temperature can reach striking values.

In our experiments mean gas temperatures of 6100 K and 6700 K have been reached [3], [4]. In this case the thermal excitation of atoms brought into the arc has to be considered.

Direct and indirect heating of the sample present in the crater of the anode:

As has been described before the arc begins to strike on the sample after indirect heating via the anode material (K-method). This is not the case when the Q-method is applied. Here the direct heating causes a less smooth vaporization, resulting in a less accurate determination of impurities.

1.2. Structure of an arc; inhomogeneous temperature

The central part of the arc (the core or column) shows the highest temperature (mean temperature 6100 K); it is surrounded by the mantle (mean temperature 4000 K) and around the mantle there is a sea of flames of perceptibly lower temperature. The arc is heated from the inside and a homogeneous temperature of the whole arc can not be expected. The conception of blowing the mantle away by a surrounding stream of nitrogen has been mentioned [6], as it appeared to be possible to suppress sodium I emission almost completely in this way, because such emission originates from the mantle [5].

The magnitude of the measuring volume in core and mantle differ greatly. By placing metal strips of increasing breadth (2-12 mm) —see also [5]—in the optical pathway near the arc it was possible to determine these magnitudes by measuring the intensities of BaI and BaII lines, in spectra obtained in this way. It has been found that the radiating volume in the mantle is three times as large as in the core.

If elements showing a low value of the ionization energy are brought into the arc, under conditions described in the introduction, then a decrease in the temperature of the core takes place because of the elements being strongly excited and emitting a strong radiation which results in much energy being withdrawn from the arc. In this way, it has been shown that in our K-arc, the original core temperature of 6100 K has been decreased to 5600 K during evaporation of sodium ($V_i \sim 5$ eV) and to 5800 K by the introduction of calcium ($V_i \sim$ 6 eV). Presumably because of the presence of relatively large amounts of carbon and carbon compounds (large diameter of the electrodes; see [7]) the concentration of elements which are being vaporized in unit time is never high and the influence of elements with a $V_i > 7$ eV could not be detected. See Fig. 2.

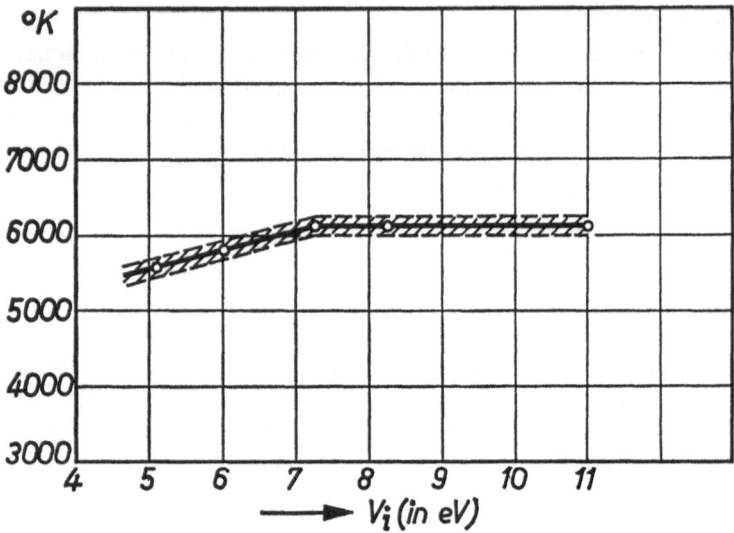

Fig. 2. Temperature of the core of the K-arc plotted as a function of the ionization potential V_i of the evaporating element.

The effect of a decrease of temperature caused by elements with a $V_i < 7$ eV has to be taken into account when other elements evaporating at the same time are brought into the arc. See further Section 2.4.

1.3. Chemical reactions taking place during heating of the analysis material placed in the anode

It will be understood, that in an environment of carbon, reduction will take place in many cases (for instance Fe_2O_3 will be converted to Fe), or even carbides will be formed (SiC from Si or SiO_2, W_2C or WC from W or WO_3, etc., see also [1]).

On the other hand, because of the fact that the arc burns in air, oxidation can also take place. For example, the heat of formation of Al_2O_3 from its components is so great that during heating in the anode crater, part of the aluminium is converted to the oxide.

Some data concerning this kind of reaction, as well as melting points and boiling points of elements, oxides and carbides have been gathered from [1, 36, 37 and 38] into Table 1.

Why should such chemical reactions be discussed here? As has been suggested already in the Introduction, the boiling point (or the temperature near to it) is of great importance to the way in which materials are vaporized. As will be pointed out in Section 1.4, the boiling point determines the speed of evaporation and the speed of introducing vapour into the arc. Volatile material has a high speed and the time spent in the measuring volume is short. Refractory material evaporates during a longer time, but its speed of evaporation is low and its atoms remain a longer time in the measuring volume; thus the chance of thermal excitation of its atoms is much greater. It will be seen later (Sections 1.4 and 2.6.1, Table 11) that the rate of evaporation is at least of equal importance as a change of temperature or a shifting of the atom-ion equilibrium.

Some factors affecting chemical reactions in the crater are listed below:

Tungsten forms stable carbides during heating; melting and boiling points remaining about the same. However, when WO_3 is heated in the crater of an anode (diameter 10 mm) the carbide formation, starting at 800 °C, is completed at 1400 °C during arcing. During this heating period, sublimation of relatively volatile WO_3 takes place. These molecules are atomized at the temperature of the core and are thermally excited. Spectra taken from WO_3 show more and stronger spectral lines.

Magnesium metal should never be analysed as such in the arc. Its reaction with oxygen takes place under emission of continuous light (flash bulbs!) originating from glowing oxide particles. Result: a strong background of the whole spectrum. It is better to convert the metal into its nitrate, and then to heat. In this way the oxide formed can be analysed easily.

What happens during heating of MgO and graphite? At 2200 °C reduction takes place and at this temperature (but not at the boiling point of the metal—1100 °C) Mg as a vapour is brought into the measuring volume.

Mg-alloys (for instance Al-Mg) on the other hand can be analysed without chemical pre-treatment. During heating in the anode, oxidation to MgO, followed by reduction at 2200 °C occurs.

Table 1

Melting points (m.p.) and boiling points (b.p.) of elements, their oxides and carbides; reaction temperatures of the oxides with C(O); decomposition temperatures (*d*) and sublimation (*s*). Temperatures are given in °C. (Data found in the literature differ greatly and are therefore not always accurate.) Many new data have been taken from [1].

Element		Oxide		Carbide		Reduction of the oxide by C(O)
m.p.	b.p.	m.p.	b.p.	m.p.	b.p.	
Ag 960	~1950	300*d*		>1400*d*		250: Ag
Al 660	~2060	2045	3500			
As	600*s*	~300	~460			low temp.: As
Au 1060	~2600					250: Au
B 2300	2550*s*	~460	~2750	2750	>3500	1600; carbide
Ba 725	1140	1920	~2000			
Be 1280	2970	2550	~3900			
Bi 270	~1470	820	1890			300: Bi
Ca 840	1490	2580	2850	~2300		
Cd 320	770	900*d*	1560*s*			850: Cd
Ce 795	1400	~2600		2540		
Co 1495	2900	1935		2300*d*		900: Co
Cr 1890	~2200	~1990	~4000	1890		1500: Cr_3C_2
Cs 30	670	400*d*				
Cu 1080	~2310	~1030*d*				400: Cu
Dy 1410	~2600	2340				
Er 1500	~2900	high				
Eu 830		2240				
Fe 1535	3100	1565		(1837)		1000?
Ga 30	~2000	1900				
Gd 1310	~3000					
Ge 940	2700*s*	1115			3700	650: Ge, GeC
Hf 2150	>3200	2810	~4300	~3890		1550
Hg −39	360	500*d*				
In 157	2000	850*s*				750?: In
Ir 2410	~4400					
K 64	770	360	1300			no reaction
La 920	3470	2315	4200			
Li 180	1320		~1700			no reaction
Mg 650	1100	2800	~3600			2200: Mg
Mn 1240	~1900	1650		(1245)		1670: carbide
Mo 2610	3700	795	1155*s*	~2700	~4500	500 → 900
	(4800)					(carbide)
Na 98	890		1300	(700)		no reaction
Nb 2470	~3700	~1500		~3500		1150
Nd 1020	3030	~1900	>4200	*d*		

Element		Oxide		Carbide		Reduction of the oxide by C(O)
m.p.	b.p.	m.p.	b.p.	m.p.	b.p.	
Ni 1450	~2900	2000		2100d		450: Ni
Os 3000	>5300	~500				500: Os
P 44–590	280	580	300s			
Pb 327	~1600	888	~2200			415: Pb
Pd 1550	~2540	750d				
Pr 935	3130	high; d		d		
Pt 1770	~4300	550d				
Rb 39	690	400d				
Re 3180	~5500	~300d				
Rh 1970	>2500	1125d				
Ru 2250	>2700	d				
Sb 630	~1500	660	1550s			
Sc 1540	~2400					
Se 220	700	345s				
Si 1410	2360	1710	~1880	~2700	~3800	1800: SiC and oxide
Sm 1070	1900					
Sn 230	2270	1130	1850s			800?: Sn
Sr 770	1300	2430	~2600	>1700		
Ta 3000	>4100	1500d		3880	5500	1200
Te 450	1390	730	1245			
Th~1700	~4000	3050	4400	2650	~5000?	
Ti 1675	>3000	1640d		3140	~4300	1300: TiC red.: Tl
Tl 300	1460	300	1080d			
U 1130	3820	~2500		~2375	~4100	
V 1890	~3000	1970		2810	3900	1100
W 3410	5930	1470s		2870	6000	800–1400: carbide
Y 1490	~2900	2410	high			
Zn 420	907	1975				900: Zn
Zr 1850	3580	2700	~4300	3540	5100	1500: ZrC

Silver and gold when heated during arcing show spluttering. This effect causes loss of material to be analysed but the analysis results cannot be trusted. The addition of, for example, nickel with which these metals form alloys or the choice of another shape of the anode prevents spluttering of these elements.

Beryllium oxide does not react with C in moist air; the metal itself, however, forms its oxide and carbide [8].

Germanium has a boiling point of almost 2700 °C. If the element is vaporized through an arc, from the crater of the anode, evaporation takes place at 3500 °C, determined according to Eq. (2), Section 1.4.3. X-ray diffraction of incompletely evaporated material shows unidentified lines (of GeC?).

If Ge is present in SiO_2 evaporation takes place at 3500 °C (formation of GeC (?) and SiC; see also [1]); carbide formation of Ge is prevented by the addition of $CaCO_3$ (CaO). In this case evaporation takes place at 2700 °C.

Lead has a boiling point near 1620 °C; the oxide boils at 2200 °C (see Table 1). At 400 °C reduction of PbO takes place. One may expect that PbO (after reduction) and Pb evaporate at the same temperature as the burnt spot of the anode in the arc. However, this appears not to be the case. If PbO is present as an impurity in substances which can be easily reduced, it takes part in the process of reduction and evaporates as lead. If it is present in a matrix, which can be oxidized even in the carbon arc (substances like Al, Ca, Ag and their compounds) PbO evaporates as such (at 2200 °C). This temperature has been determined via S_t (see Eq. (2), p. 15).

Thermal decomposition of salts, carbide formation. Sulphates, phosphates, carbonates, nitrates, etc., are converted into the oxides and the metal oxides thus formed can evaporate as such or as metals formed by reduction with C, CO or as carbides. Nickel [1], Schroll [40] and Rautschke [2] have examined by thermodynamic means and by X-ray diffraction when and how metals and carbides are formed in the d.c. carbon arc. They distinguish two main groups:

  (i) stable carbides of B, Si, Ti, Zr, Hf, V, Nb, Ta, Cr, Mo and W;
  (ii) unstable carbides of Mn, Fe, Co and Ni.

The carbides of Al, U, Th, the rare earths in general and the alkaline earths Ca, Sr, Ba, Be and Mg are highly sensitive to air and moisture due to their saline nature, see [1].

Table 1 contains the melting and boiling points of the elements and their oxides and also some data concerning their decomposition and reactions with C or CO.

1.4. Evaporation of the analysis material or of its reaction products formed during arcing

1.4.1. It has been pointed out in the Introduction, that the boiling points of elements or compounds present in the sample to be analysed are of major importance, certainly of the same importance as the temperature of excitation or the shifting of atom-ion equilibria. The longer the residence time of the atoms in question, in the radiating volume, the higher the intensity of spectral lines originating from the excited atoms. Volatile substances travel through the arc more quickly than non-volatile ones as the latter evaporate at a lower speed during a longer time. In the case of volatile substances, one may also think of larger lateral losses particularly if the Q-arc be considered. The explanation of this (see Fig. 3) is that the temperature of the anode tip, approximately that of the sample, increases with the exposure time according to curve c. For low boiling substances

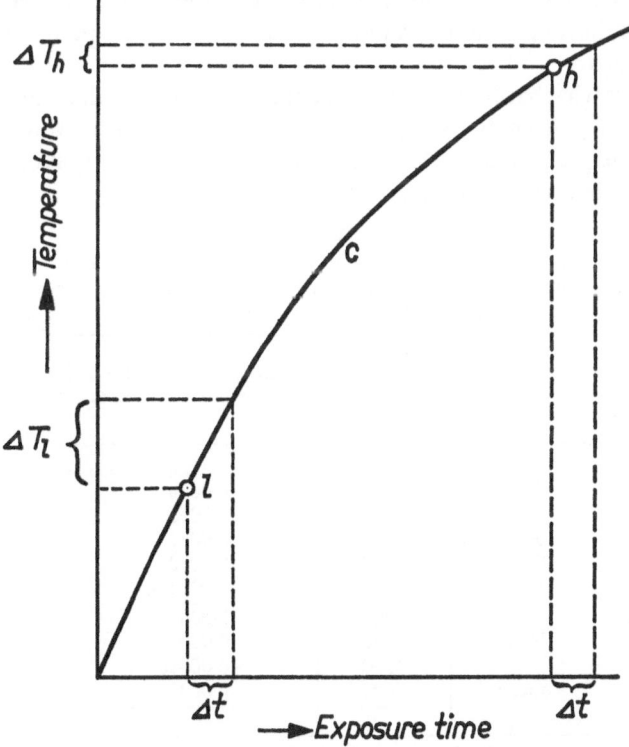

Fig. 3. On the abscissa the exposure time (up to 5 minutes); on the ordinate (up to 2500 °C) the temperature of the tip of the anode or of the sample. In a short interval of time (Δt) the tip temperature has risen to a much higher temperature in case of low-boiling substances (ΔT_1 relatively large) than in case of high boiling refractory materials (ΔT_h relatively small).

(boiling point: l) the temperature gradient is large, whilst for high boiling substances (b.p.: h) the gradient is small. This means that the sample with the low b.p. evaporates not only at its b.p. but also at a temperature considerably higher, resulting in a more or less explosive way of evaporation. This is much less the case with high boiling substances. Compare the magnitudes ΔT_1 and ΔT_h for the same interval of time Δt.

It can be repeated that, in this procedure of complete evaporation, there is no question of pre-burn, i.e. of waiting for a state of constant evaporation before exposing. Pre-burn always causes a loss of sensitivity in volatile elements or compounds! In the case of pre-burn, nobody is concerned with the speed of introducing a sample into the arc, in contrast to the careful regulation of the gas speed in flame photometry.

From what has been said above, concerning the time that the atoms originating from the sample remain in the measuring volume, it can be expected that the limits of detection of high boiling substances are low and vice versa (see Table 2 and [9]).

Table 2

Limits of detection of impurities (in mg) present in a 5 mg sample determined according to the K-method for elements showing a $V_i = 9 - 11$ eV.

Element	b.p. (°C)	Limit of Detection (in mg) experimentally found
As	600	0·0050
Cd	770	0·0015
P	450	0·0037
Zn	900	0·0010
mean	~680 °C	~0·0028 mg
Au	2600	0·00017
Be(O)	3900	0·000005
Pt	4300	~0·00030
mean	~3600 °C	~0·00016 mg

The conclusion is that *Low boiling elements show a high limit of detection, high boiling elements show a low limit provided that they are vaporized completely*; differences of V_i, etc., have not been taken into account.

1.4.2. Visual observation, during the evaporation of a sample through the arc, can be demonstrated by a soda lime glass. During the temperature rise of the anode tip, according to curve c (Fig. 3), sodium begins to evaporate at about 1300 °C (b.p. of Na_2O or NaOH). The well-known yellow-orange colour of Na is observed. During a further increase of temperature, SiO_2 (b.p. $\sim$2000 °C) starts to vaporize. At this point there is no typical colour of the arc and no difference with a C–C discharge.

If the temperature of the sample is increased to about 2800 °C, evaporation of CaO begins and the arc becomes reddish coloured. By taking time-resolved spectra, control of these visual observations can be verified.

1.4.3. The speed of evaporation S_i can be determined from data found experimentally (S_i is given in an *arbitrary scale*). If 5 mg of an element evaporates from the anode through an area of 0·9 mm² and that the vapour is brought to a temperature of 6100 K (temperature of the core of the arc) then at this temperature the volume of 5 mg of the element amounts to about $(22\cdot4 \times 5 \times 200)/(9A)$ ml; (A = the atomic weight). The factor 22·4 corrects for heating $5/A$ mga up to 6100 K.

This quantity of vapour passes through the area of 0·9 mm² (0·009 cm²) in t seconds (t = time of evaporation) and the volume of vapour passing per second amount to: $(22\cdot4 \times 5 \times 200)/(9\,at)$ ml; this volume is equal to $S_i \times 0\cdot009$ ml.

Therefore:
$$S_i = \frac{280\,000}{At} \text{ cm/s relatively} \tag{1}$$

The area of 0·9 mm² has been chosen so as to obtain a value of S_i of an element showing a 'mean' boiling point (such as copper) equal to about 100 cm/s [39].

The time of evaporation can be determined by taking spectra in small steps. Table 3 shows some results of the magnitude of t and S_i calculated according to Eq. (1).

The relationship between S_i and boiling point is shown in Fig. 4 and the curve found can also be expressed by the following formula:
$$\log S_i = 2\cdot94 - 0\cdot441\,T_{bp} \times 10^{-3} \tag{2}$$
where T_{bp} denotes the boiling point (°C) of the element or compound in question.

Table 3

Element (atomic weight A in brackets)	b.p. (°C)	t in s (K-method)	S_i in cm/s (Eq. (1))
Te (128)	1390	10	220
Cu (63½)	2300	55	80
Fe (56)	3000	125	40
Mo (96)	3700	146	20
Th (232)	4400	120*	10

* Expressed in gram atoms, the quantity of Th is much less than for instance of Mo.

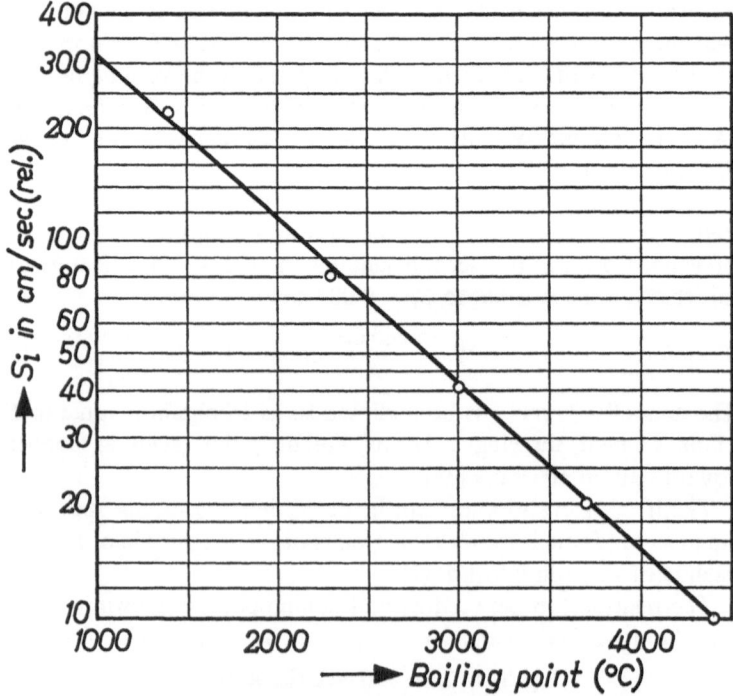

Fig. 4. The speed of evaporation S_i of elements or compounds logarithmically plotted against linearly their boiling points. The higher the boiling point, the smaller the speed of evaporation.

An example of calculating ratios of S_i of different metals via their vapour pressures is given in Section 2.6.2.

Conclusion

(1) The straight curve given in Fig. 4 and formulated by equation (2) is based on heating an anode of 8–10 mm diameter by a d.c. discharge of 10 A. If the diameter of the anode tip is decreased, the temperature curve (see Fig. 3) will be steeper at the beginning of exposure. Later on, with substances of high boiling points, temperature gradients will show no difference.

(2) According to [10] a 10 A arc causes a heat development of about 250 W/s per electrode, whereas the heat of combustion of C in the air (1–1½ mg/s) delivers about 40 W. Both values taken together deliver a supply of 70 cal/s and cause a maximum temperature of the anode tip to be about 3500–4000 °C, limited by radiation losses, convection and conductivity and also by the heat of evaporation of C ($\sim$3600–4000 °C). (The number of calories necessary for the evaporation of the sample amounts to $\sim$0·1 cal/s, a quantity completely negligible to the 70 cal/s mentioned above.)

(3) The ratio of limits of detection for 'mean boiling points' of 680 °C and 3600 °C as mentioned in Table 2 amounts to 0·0028/0·00016 = 18. The ratio of 'mean speeds of evaporation' at the same temperature amounts to 436/23 = 19, according to Fig. 4 and equation (2), an agreement which may be called reasonable in view of the fact that excitation (and ionization) potentials differ.

1.4.4. *An example of the application of the magnitude of S_i in spectral analysis*

The speeds of evaporation of Pb and PbO have been found to be 180 and 85 rel. cm/s respectively, corresponding to boiling points of 1600 and 2200 °C. The ratio of both values of S_i amounts to 180/85 = $\sim$2 and this factor has to be applied in the following way:

The K-factors of various Pb lines have been determined by analysing metallic samples. If oxide samples are analysed (S_i is low, time of residence of the atoms in the measuring volume is relatively longer and intensities of spectral lines are high) results have to be multiplied by 1/2 = 0·5.

It is not necessary to memorize these correction factors, if necessary, they are mentioned for each element in Section 4.2.2.

1.5. Evaporation and Atomization

As the method of analysis described in this book has been aimed at analyses of all kinds of materials, all the processes which can take place during arcing will be described here. As an example, take $BaSO_4$. During heating in the crater, it reacts with C, and BaS as an intermediate product is formed (tested by adding some acid to the remaining substance after interruption of the arc; smell of H_2S). After further heating BaO remains. At about 1200 °C BaC_2 is formed, but at higher temperature the carbide decomposes and because of the presence of moisture and oxygen (the arc 'burns' in air) BaO if formed (see Section 1.3), thermal decomposition of salts and carbide formation. When the boiling point of BaO (about 2000 °C) is reached, Ba evaporates as BaO with a speed of $S_i = 115$ cm/s and not as Ba at its boiling point of 1140 °C with a speed of $S_i = 275$ cm/s.

The oxide molecules enter the arc, that is, they are brought to the mean temperatures of both core and mantle. Now we have to consider the following equilibrium:

$$BaO \leftrightharpoons Ba + O*$$

Following the considerations given in [11] about 40% of the BaO molecules in the mantle and 90% in the core are dissociated.

The Ba atoms thus formed only partially participate in thermal excitation and emission, because of the following equilibrium ([20], see also [11]):

$$Ba \leftrightharpoons Ba^+ + e^-*$$

* For both equilibria the following equations hold:

dissociation	ionization
$MO \leftrightharpoons M + O$	$M \leftrightharpoons M^+ + e^-$
degree of dissociation D_{ox}:	degree of ionization α:
$D_{ox} = \dfrac{[M]}{[MO] + [O]}$	$\alpha = \dfrac{[M^+]}{[M] + [e^-]}$
dissociation constant K_T:	ionization constant K':
$K_T = \dfrac{[M] \times [O]}{[MO]}$	$K' = \dfrac{[M^+] \times [e^-]}{[M]}$
K_T is also (:) exp. $(-E/kT)$.	K' is also (:) exp. $(-V_i/kT)$.
E is dissociation energy;	V_i is ionization energy;
the greater E, the smaller K_T,	the greater the V_i, the smaller K',
$[M]$, $[O]$ and D_{ox}	$[M^+]$, $[e]$ and α

As a rule of thumb it may be said that a temperature of about 3000 K will just bring about a measurable dissociation or ionization (and also excitation) of $4\frac{1}{2}$–5 eV, while at a temperature of 4500 K this value amounts to about 7 eV and at 6000 K some 10 eV.

In consequence of its relatively low ionization potential (V_i Ba = 5·2 eV) ionization is appreciable: in the mantle 50% of the Ba atoms remain as such and in the core only 0·2%.

From the original BaO molecules $\frac{1}{2}$ of the 40% (i.e. 20%) remains as Ba atoms in the measuring volume of the mantle, and $\frac{1}{5}$ of the 90% (i.e. 18%) remains in the core. Both percentages give rise to the emission of BaI lines. For further considerations, see Section 1.6.

As has been mentioned before, many elements either in the oxide form or as such do not give rise to such complicated processes as the oxides are reduced and evaporation takes place at the boiling point of the element itself. Further on, shifting of atom-ion equilibria becomes less if one moves in the periodic system from left to right. As may be seen from [11] many oxides are completely dissociated in the core of the arc. From values of chemical thermodynamic properties [37] it can be deduced, although very approximately, that the standard heat of formation of compounds (oxides) increases linearly with their melting points; also, the higher the melting point, the more stable the oxide.

1.6. Origin of the emitted light; from core or mantle?

1.6.1. The alkali metals ($V_i \leqslant 5\cdot4$ eV) are highly ionized at the temperature of the core (6100 K), but less ionized at 4000 K (temperature

Table 4

Degree of Ionization of the Alkalis, the Alkaline Earths, Be and Mg (after [12])

Element	V_i in eV	Degree of ionization (in %)		Remaining atoms (in %)	
		at 6100 K	at 4000 K	at 6100 K	at 4000 K
Li	5·4	99·6	25	0·4	75
Na	5·1	99·8	48	0·2	52
K	4·3	100	94	0·0	6
Rb	4·2	100	96	0·0	4
Cs	3·9	100	100	0·0	0
Be	9·3	16	0	84	100
Mg	7·6	81	0	19	100
Ca	6·1	98·6	9	1·4	91
Sr	5·7	99·3	20	0·7	80
Ba	5·2	99·7	40	0·3	60

of the mantle). In both cases the remaining atoms are thermally excited and give rise to emission of discrete wavelengths. Table 4 shows clearly that arc lines (I lines or atom lines) originate mostly from the remaining atoms present in the mantle and not from the core.

It is further shown in Table 4 that the alkaline earths behave more or less the same as the alkali metals, but elements with a $V_i > 7$ eV (Be and Mg) show a degree of ionization at 6100 K which is sufficiently small to produce emission from the core. Therefore it is necessary to know the ratio N_e/N_0 (N_e = the number of excited atoms) equal to part of all the atoms available (N_0) which is excited thermally and which is a measure for the intensity or energy of the emitted light (Maxwell/Boltzmann's partition law). The ratio N_e/N_0 is proportional to exp. $(-V_e/kT)$; V_e = the excitation potential of the spectral line under consideration, in this case a resonance line; k = Boltzmann's constant and T = the absolute temperature of excitation.

1.6.2. The results of the calculation concerning the product of remaining atoms (after ionization) and the part of them which is excited according to equation (3) is shown in Table 5. The products which are mentioned in Table 5 are a measure of the intensity of the resonance lines emitted. However, we have to consider that measuring volumes of core and mantle are in the proportion of 1:3 (see Section 1.2), which means that products mentioned under mantle excitation (Table 5) have to be multiplied by three. After correction by this factor it is now possible to calculate the percentage of the emitted light of resonance lines originating from the mantle and core (see Table 6).

Results mentioned in Table 6 are shown graphically in Fig. 5. Resonance arc lines of elements with a $V_i = 7.3$ eV originate for 50% from the core and for 50% from the mantle of the arc; if $V_i > 7.3$ eV, corresponding light is emitted chiefly from the core and for $V_i < 7.3$ eV, resonance lines originate mainly from the mantle.

1.6.3. Up to this point we have discussed the origin of atom resonance lines. We started from atoms remaining after ionization. Let us now consider the emission of resonance ion lines. A similar calculation for these lines gives results as shown in Table 7.

Table 5
Calculation of the Product of Remaining Atoms

Element	V_i (in eV)	Mantle excitation V_e (in eV)	Atoms available (see Table 4)	N_e/N_o	Product $\times 10^7$	Core excitation Atoms available (see Table 4)	N_e/N_o	Product $\times 10^7$
Be	9·3	5·25	1·00	$2\cdot35 \times 10^{-7}$	2·3	0·84	$3\cdot72 \times 10^{-5}$	313
Mg	7·6	4·41	1·00	$2\cdot63 \times 10^{-6}$	26·3	0·19	$1\cdot90 \times 10^{-4}$	361
Ca	6·1	2·92	0·91	$2\cdot00 \times 10^{-4}$	1820	0·014	$3\cdot47 \times 10^{-3}$	485
Sr	5·7	2·68	0·80	$4\cdot07 \times 10^{-4}$	3250	0·007	$6\cdot61 \times 10^{-3}$	463
Ba	5·2	2·23	0·60	$1\cdot51 \times 10^{-3}$	9060	0·003	$1\cdot32 \times 10^{-2}$	396

Table 6
Contribution of the Emitted Light (Resonance arc Lines)

Element	V_i (in eV)	From the mantle ($\times 10^7$)	From the core ($\times 10^7$)	From mantle plus core ($\times 10^7$)	Percentage from: Mantle	Core
Be	9·3	7	313	320	2%	98%
Mg	7·6	85	361	446	19%	81%
Ca	6·1	5460	485	5945	92%	8%
Sr	5·7	9750	463	10213	95½%	4½%
Ba	5·2	27180	396	27576	98½%	1½%

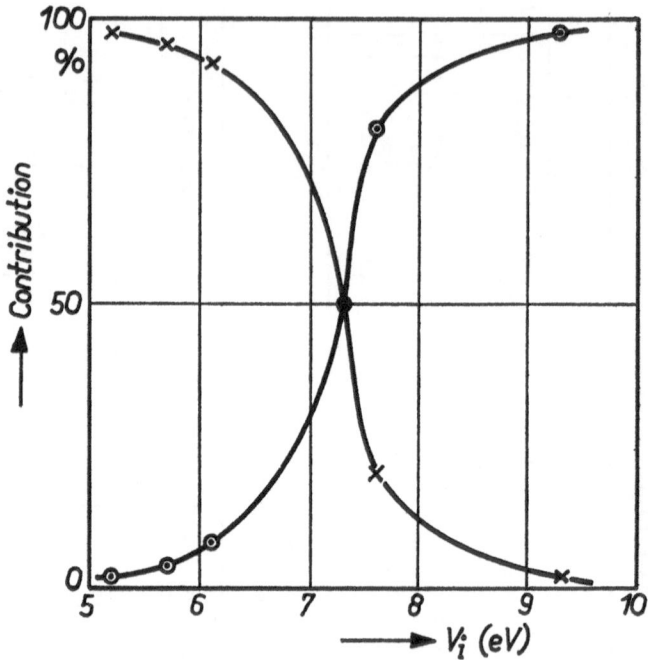

Fig. 5. Contribution of the energy by resonance arc lines emitted from the mantle (×) and from the core (○) of a d.c. K-arc in dependence on the ionization energy.

Table 7

Contribution of the Emitted Light (Resonance Spark Lines)

Element	V_i (in eV)	V_e (in eV)	From the mantle	From the core
Be	9·3	3·94	0%	100%
Mg	7·6	4·41	0%	100%
Ca	6·1	3·14	0·5%	99·5%
Sr	5·7	3·03	1·1%	98·9%
Ba	5·2	2·71	2·9%	97·1%

From this table it is clear that all ion lines originate from the core.

1.6.4. In conclusion, we can ascertain the following facts:

    a. Atom lines of elements with a $V_i < 7·3$ eV are emitted by the mantle;

b. Atom lines of elements with a $V_i > 7\cdot3$ eV are chiefly emitted by the core;

c. Ion lines of all elements originate from the core.

As the core is least influenced by the surrounding atmosphere, emission originating from the core delivers more accurate results than mantle emission.

Chapter 2

Excitation and Emission

In order to obtain quantitative results from arc analysis, one must measure line intensities in spectra, but before doing this it is advisable to consider the extent to which the luminous vapour of the arc resembles the original sample in its composition. It is necessary to examine the formulae of the relationships which govern the entry of elements into the excited state in the core of the arc. These formulae are briefly stated in Section 2.1 and they permit the derivation of factors called K-factors (and Q-factors) which are then used to work out numerical values of the analyses. It should be understood that these are not two independent variables, but are factors related to two different ways of photographing part of the radiation from the arc, which are appropriate for different samples. The factors are defined in Section 2.2.

The twofold layer structure of the arc, with its hotter core and cooler mantle, has the result that some radiation emitted from the core may be absorbed in passing through the mantle, and this causes a selective weakening of some spectral lines, which is discussed in Section 2.3. An analyst will be disappointed if he expects to eliminate the influence of chemistry on his problems by changing from a 'wet' to a 'physical' method of analysis, from reactions in solutions to high temperature reactions. As was said in Section 1.4, the rate of evaporation of an element is affected by other elements in the crater with it, this is called the matrix effect, whether it is due to natural minerals, allows, or to diluents which have been added to the sample in the arc crater. The physical process of evaporation and ionization is preceded by chemical reactions and interactions between moving vapours which sometimes have to be treated empirically; and these are discussed in Sections 2.4, 2.5 and 2.6. Matrix effects are not all interferences with analysis, some are of use in controlling the rate of evaporation of volatile elements or of smoothing out variations in arc core temperature which might otherwise exaggerate the variability of the arc.

2.1. Thermal excitation; formulae (see also [13] and [5])

The general expression for the intensity I of a spectral line is:

$$I = N_e a h \nu \tag{4}$$

where N_e = the number of excited atoms (ions) present in the measuring volume;

a = the transition probability; $1/a$ = the lifetime of the excited state in seconds;

h = Planck's constant = $6 \cdot 6 \times 10^{-27}$ erg s;

ν = frequency; $c/\nu = \lambda$, expressed in cm.

Maxwell/Boltzmann's partition law gives us the relationship N_e/N_0, N_0 being the total number of atoms (ions) present in the measuring volume:

$$N_e/N_0 = g\{\exp. (-V_e/kT)\} \times 1/B \tag{5}$$

Here g = the statistical weight factor of the energy level under examination $(g = 2J + 1$; J = the internal quantum number; see [14]);

V_e = the excitation energy in eV (1 eV = $1 \cdot 6 \times 10^{-12}$ erg);

k = Boltzmann's constant ($= 1 \cdot 37 \times 10^{-16}$ erg/°C);

T = the temperature of excitation in K;

B = the partition function (for numerical values see [12] and [15]).

Combining Eq. (4) and (5) we find:

$$I = N_0 g a h \nu \{\exp. (-V_e/kT)\} \times 1/B \tag{6}$$

This equation contains numerical factors with the exception of I and $h\nu$, which indicate a dimension of an energy. The intensity, I, has to be placed in an arbitrary scale equal to unity, which means a density of $0 \cdot 07$ on the H and D curve (Section 3.2, Fig. 10). In fact an energy many times greater has been emitted by the light source and this ratio we may call the 'instrumental factor', f. This factor f has been determined by measuring the energy of four Cd-lines and applying ga-values mentioned in [15] and [16]. The degree of ionization of Cd at 6100 K amounts to $0 \cdot 764$ and f has been found to be $1 \cdot 09 \times 10^8$ (optical positioning, Hilger large quartz E 492, K-method, Section 4.1.1, Fig. 14).

Anticipating on the definition of K-factors (Section 2.2) the following equation holds for atoms present in the test volume:

$$N_0 = \{0.01\ K \times 0.005 \times 6 \times 10^{23} \times 1/p_{\text{rel}} \times (1 - \alpha) \times D_{\text{ox}}xv_m\}$$
$$/(A \times S_t) = 3K \times (1 - \alpha) \times D_{\text{ox}}$$
$$\times 1/p_{\text{rel}} \times v_m \times 10^{19}/(A \times S_t) \qquad (7)$$

where $K = K$-factor (Section 2.2)

α = degree of ionization

D_{ox} = degree of dissociation of oxide molecules formed in the arc (see Section 1.5)

$1/p_{\text{rel}}$ = reciprocal of the relative sensitivity of the photographic plate dependent on the wavelength [3]

v_m = ratio of volumes measured (Section 1.2 and [5])

A = atomic weight of the element vaporized

S_t = speed of evaporation (see Fig. 4 and Eq. (2)).

Combination of equations (6) and (7) delivers after having put I (in Eq. (6)) equal to 1.09×10^8:

$$1.09 \times 10^8 =$$
$$[3\ K(1 - \alpha) \times D_{\text{ox}} \times 1/p_{\text{rel}} \times v_m \times 10^{19} \times gahv$$
$$\{\exp.\ (-V_e/kT)\}]/(AS_tB)$$

or in numerical terms and after taking logarithms:

$$\log a = [f_T V_e + \log \lambda + \log A + \log S_t + \log B] -$$
$$[5.75 + \log K + \log (1 - \alpha) +$$
$$\log D_{\text{ox}} + \log 1/p_{\text{rel}} + \log g + \log v_m] \qquad (8)$$

where f_T = a factor which equals 0.833 if the excitation takes place at 6100 K, 0.893 at 5800 K and 1.25 at 5600 K.

Note. For ion lines α instead of $(1 - \alpha)$ has to be applied.

Equation (8) has been given here, because it has been shown in [5] that all factors influencing the excitation process have been taken into account. The mean ratio of A-values calculated according to Eq. (8) and of those mentioned in [15] appeared to be 0.75, a figure which may be called satisfying in view of the numerous factors influencing the whole process (cf. Table 8). The largest single source of error is the uncertainty in the degree of ionization (after Corliss and Bozman [15]).

At the end of this chapter it is worth while to quote the following sentences [17], the content of which being also already applied in

Sections 1.5 and 1.6: 'It is correct to bear the ionization energy in mind when investigating spark lines (compare however [18]), but it must be considered in the same way as Ahrens considers it [19], where, according to Saha's equation [12, 20], for a given temperature the ionization energy is a measure of the degree of ionization. Boltzmann's equation can then be applied to the number of ions determined in this way, this giving the ratio of the excited ions and the total number of ions at a given temperature.'

2.2. Definition of K- and Q-factors and their relationship

In general the following equation holds:

$$C = K \times I \qquad (9)$$

where C = the concentration of the element (in weight %) under consideration present in the sample to be analysed; mass of the sample = 5 mg;

I = the intensity of one of the spectral lines of the element under consideration;

K = a proportionality factor, equal to C if I = unity.

One may also say that the spectrochemical definition of the K-factor for a certain wavelength of an element states, that K% of 5 mg of the material to be analysed will cause the spectral line concerned to have the relative unit intensity on the photographic plate. A relative unit intensity means a density of 0·07 on the H and D curve (see further Section 3.2, Fig. 10).

Equation (9) concerns the so-called K-method (see Introduction). Exactly the same considerations hold for the Q-method:

$$C = Q \times I \qquad (10)$$

Here C = the concentration of the element under consideration, but the mass of the sample amounts to 10 mg.

In [4] a derivation is given of the relationship K/Q and V_i.

For emission lines *originating from the core* the following equation holds (by approximation):

$$\log (K/Q) = 0.24\, V_i - 0.64 \qquad (11)$$

For those *originating from the mantle*:

$$\log (K/Q) = 0.042\, V_i + 0.86 \qquad (12)$$

Table 8
Calculation of transition probabilities from K-factors

Element V_i (eV) T (K)	λ (Å)	V_e (eV)	Evaporation as	S_i	A	B	K in %	1 − α	D_{ox}	v_m	g	$1/p_{rel}$	a_{calc} $10^8 \times$	$a_{C\&B}$ $10^8 \times$	$a_{calc}/a_{C\&B}$
Au I 9·2 6100	2675·95	4·6	Au (2600)	65	197	2·68	·01	·80	1	1	2	1·18	0·59	0·55	1·07
Cd I 9·0 6100	3261·06	3·78	CdS	110	112	1·00	·1	·764	1	1	3	1·0	0·0044 (Allen)	0·003 / 0·0045	(1·45)
Co I 7·9 6100	3453·50	4·0	Co (2900)	40	58·9	38	·008	·32	1	1	12	0·80	0·48 (Allen)	2·2 (1·2)	(0·22)
	3465·80	3·56	(2900)	40	58·9	38	·015	·32	1	1	12	0·80	0·11	0·2	0·54
	2521·36	4·89	(2900)	40	58·9	38	·019	·32	1	1	8	0·77	1·25	2·5	0·50
	3044·00	4·05	(2900)	40	58·9	38	·025	·32	1	1	10	1·11	0·135	0·31	0·43
Cr I 6·8 6100	3013·71	5·06	Cr (2200)	100	52	12·6	·2	·05	1	1	5	1·13	1·05	2·0	0·53
	2731·91	5·45		100	52	12·6	·5	·05	1	1	5	1·2	0·73	1·6	0·45
Ga I 6·0 4000	2943·64	4·29	Ga (2000)	120	69·7	5	·012	·95	1	3	6	1·15	2·1	1·8	1·18
	2874·24	4·29	(2000)	120	69·7	5	·025	·95	1	3	4	1·18	1·5	1·5	1·00
	2944·18	4·29	(2000)	120	69·7	5	·1	·95	1	3	4	1·15	0·39	0·45	0·86
	2719·65	4·64	(2000)	120	69·7	5	·3	·95	1	3	2	1·16	0·65	1·0	0·65
Ge I 8·1 5800	2651·18	4·83	GeO₂ (2700)	60	72·6	8	·008	·68	1	1	5	1·03	1·5	5·2	0·29
	3039·06	4·94	(2700)	60	72·6	8	·01	·68	1	1	3	1·13	2·0	8·0	0·25
	2709·63	4·62	(2700)	60	72·6	8	·016	·68	1	1	1	1·15	1·8	12	0·15
	2754·59	4·65	(2700)	60	72·6	8	·017	·68	1	1	3	1·15	·60	3·3	0·19
	2651·58	4·65	(2700)	60	72·6	8	·02	·68	1	1	3	1·03	·55	2·7	0·14
	2691·34	4·65	(2700)	60	72·6	8	·03	·68	1	1	3	1·18	·33	2·4	0·14
	3269·49	4·65	(2700)	60	72·6	8	·07	·68	1	1	3	0·9	·22	0·67	0·34
Hg I 10·4 6100	2536·52	4·87	Hg (360)	600	200·6	1	·02	1	1	1	3	0·8	1·3	1·2	1·10
In I 5·8 4000	2932·63	4·48	In (2000)	120	114·8	4	·27	·95	1	3	2	1·14	0·65	1·2	0·54
Mg I 7·6 6100	2852·13	4·33	MgO (1650)	~160	24·3	1·04	·0006	·19	1	1	3	1·18	2·1	3·1	0·68
Mn I 7·4 6100	2798·27	4·41	Mn (1900)	120	54·9	6·9	·004	·14	1	1	6	1·18	2·7	1·1	2·50
	2801·06	4·41	(1900)	120	54·9	6·9	·005	·14	1	1	4	1·18	3·3	1·2	2·70
	3044·57	6·2	(1900)	120	54·9	6·9	·14	·14	1	1	8	1·1	2·1	2·6	0·81
	3073·13	6·2	(1900)	120	54·9	6·9	·45	·14	1	1	4	1·09	1·3	2·7	0·50
	3070·27	6·2	(1900)	120	54·9	6·9	·6	·14	1	1	6	1·09	0·66	1·8	0·37

Element V_i (eV) T (K)	λ (Å)	V_e (eV)	Evaporation as	S_i	A	B	K in %	1 − α	D_ox	v_m	g	1/p_rel	a_calo ×10^8	a_C&B ×10^8	a_calo/a_C&B
Mo I, 7·4, 6100	3132·59	3·94	MoO₃ (800)	386	96	11·2	·05	·14	1	1	9	1·08	0·64	1·1	0·59
	3170·35	3·89	(800)	386	96	11·2	·06	·14	1	1	7	1·03	0·67	0·77	0·88
	3193·97	3·89	(800)	386	96	11·2	·063	·14	1	1	5	1·03	0·90	0·88	1·00
	3158·16	3·9	(800)	386	96	11·2	·18	·14	1	1	7	1·03	0·21	0·54	0·40
Ni I, 7·6, 6100	3002·49	4·14	Ni (2900)	45	58·7	33	·009	·19	1	1	7	1·13	1·0	0·96	1·03
	3050·82	4·07	(2900)	45	58·7	33	·01	·19	1	1	9	1·1	0·63	0·57	1·10
	3003·63	4·22	(2900)	45	58·7	33	·02	·19	1	1	5	1·13	0·72	0·90	0·80
	3315·66	3·85	(2900)	45	58·7	33	·08	·19	1	1	7	0·9	0·09	0·10	0·90
	2992·60	4·15	(2900)	45	58·7	33	·1	·19	1	1	5	1·13	0·13	0·17	0·73
	2943·91	4·22	(2900)	45	58·7	33	·1	·19	1	1	5	1·14	0·14	0·20	0·70
Pb I, 7·4, 6100	2833·06	4·36	PbO (2200)	85	207	1·9	·025	·139	1	1	3	1·17	0·59	0·60	0·99
Rh I, 7·7, 6100	3434·89	3·60	Rh (2500)	70	103	30·4	·02	·224	1	1	12	0·81	0·31	0·34	0·91
	3396·85	3·64	(2500)	70	103	30·4	·03	·224	1	1	10	0·81	0·27	0·31	0·87
Ru I, 7·5, 6100	3498·94	3·54	Ru (2700)	60	102	41·1	·03	·157	1	1	13	0·8	0·29	0·46	0·63
Sb I, 8·6, 6100	2528·52	6·10	Sb (1500)	200	122	5·2	·035	·516	1	1	4	0·8	12	14	0·86
	2598·05	5·80	(1500)	200	122	5·2	·034	·516	1	1	2	1·0	11	32	0·35
	2877·92	5·34	(1500)	200	122	5·2	·11	·516	1	1	2	1·16	1·4	2·8	0·50
Si I, 8·1, 6100	2516·11	4·93	SiO₂ (1880)	120	28·1	9·8	·006	·371	1	1	5	0·8	2·2	2·6	0·84
	2881·60	5·06	(1880)	120	28·1	9·8	·008	·371	1	1	3	1·19	2·6	5·0	0·54
	2506·90	4·93	(1880)	120	28·1	9·8	·016	·371	1	1	5	0·8	0·85	1·3	0·66
	2528·51	4·93	(1880)	120	28·1	9·8	·017	·371	1	1	3	0·8	1·4	2·4	0·57
Sn I, 7·3, 6100	2514·32	4·93	Sn (2270)	80	119	5·8	·015	·122	1	1	3	0·8	1·25	1·9	0·68
	3175·05	4·31	(2270)	80	119	5·8	·015	·122	1	1	3	1·05	2·1	1·1	1·90
	2839·99	4·77	(2270)	80	119	5·8	·018	·122	1	1	5	1·16	2·5	4·2	0·60
	3034·12	4·28	(2270)	80	119	5·8	·020	·122	1	1	1	1·12	4·5	4·4	1·00
	2863·33	4·31	(2270)	80	119	5·8	·025	·122	1	1	4	1·15	1·3	1·8	0·72
	3009·14	4·31	(2270)	80	119	5·8	·02	·122	1	1	3	1·13	1·1	1·3	0·85
V I, 6·7, 6100	3185·40	3·94	V (3000)	40	51·0	55·9	·03	·043	1	1	12	1·03	1·15	2·2	0·53
	3183·41	3·89	(3000)	40	51·0	55·9	·03	·043	1	1	8	1·03	1·05	2·5	0·42
Zn I, 9·4, 6100	3075·90	4·01	Zn (900)	350	65·4	1	·25	·9	1	1	3	1·1	·0037	·0043	0·85
														mean:	0·754

Both formulae are best calculated for transitions situated not high in the term scheme (for instance resonance lines and some more). They enable *calculating Q-factors from K-factors and vice-versa*, which is of great value to the analyst.

For which value of V_i are K/Q equal? If this ratio is calculated by Eq. (11) and Eq. (12)? By putting $0.24\ V_i - 0.64 = 0.042\ V_i + 0.86$, we find $V_i = 7.5$ eV, which agrees well with $V_i = 7.3$ eV found in Section 1.6.2, Fig. 5, where it is shown that at this value of V_i 50% of the emitted energy originates from the core and 50% from the mantle.

2.3. Self-absorption causes gradually increasing K- and Q-factors

As stated in Eq. (9) a linear relationship exists between the concentration, C, and the intensity, I, originating from thermally excited atoms. In cases where the emitted light is absorbed by other atoms of the same kind (mostly in the ground state) re-emission from the latter takes place in *all* directions, which causes a loss of energy. The more atoms present, the greater the decrease of intensity (see Fig. 6a). If C and I are plotted in a double-logarithmic scale, self-absorption is characterized by $\tan \alpha'' < 1$ and $\alpha'' < 45°$, whereas in case of no self-absorption $\tan \alpha' = 1$ and $\alpha' = 45°$ (see Fig. 6b).

For self-absorption Eq. (9) no longer holds, but must be replaced by

$$C = K \times I^n; \tag{13}$$

where n must be >1 in order to compensate for loss of emitted energy.

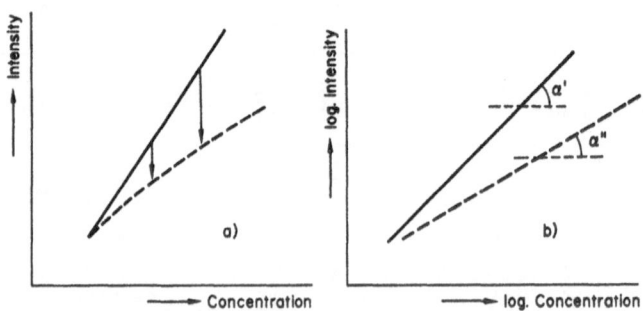

Fig. 6a and b. Concentration of atoms present in the radiating volume of the arc versus the intensity of a spectral line originating from these atoms. The higher the concentration the larger the absorbed energy (Fig. 6a) and the smaller α'' (Fig. 6b).

In the practice of spectrochemical analysis, concentration values are not determined after Eq. (12), but the range of increasing values of K is utilized and once and for all determined.

An example may be given (Table 9). The copper line 3247·54 I, showing self-absorption (tan $\alpha'' = 0·85$), can be easily used for calculating concentration values by applying K-factors increasing with increasing intensity.

Table 9

Because of self-absorption of the Cu 3247 I line, K-factors increase with increasing intensity.

Intensity	K-factor
1	·0004%
5	·0005
10	·0006
(20)	(·0007)
(40)	(·0008)

Note. It is not advisable to measure intensities higher than 10, because of line-broadening which is not taken into account if only the centre of the line is measured.

Self-absorption of spectral lines can be expected where the excitation of the atom takes place from the ground level. The atom is excited up to a certain level, from which it falls back under emission of light of discrete wavelength. This light can be absorbed and re-emitted as described above. According to [14] the Cu-emission just mentioned has been characterized as follows: low level 0·00 eV; high level 3·80 eV. The low level equal to 0·00 eV means the ground state of the copper atom.

2.4. Matrix effects. Change of temperature of excitation (application of Maxwell–Boltzmann's distribution law)

As will be explained in Section 2.6, matrix influences caused by a change of excitation temperature (this section) or by a shifting of atom/ion equilibria (Section 2.5) or by a change of the speed of evaporation (Section 2.6) are more or less of the same order. In the majority of cases mentioned in the literature, a change of the speed evaporation has been omitted from the discussion.

In Section 1.2, mention is made of the alkalis decreasing the temperature of the core from 6100 K down to 5600 K. If an element enters the core and is brought to excitation *at the same time*, the emitted energy will be less and in order to determine the real concentration of this element, a factor of >1 has to be applied. An example will be described illustrating how to solve such problems.

Zinc in NaOH evaporates at approximately the same time as sodium in NaOH, but not exactly, as the boiling point of zinc is about 900 °C and NaOH boils at about 1400 °C (experience has shown that, at a difference of boiling points of *about* 1000 °C or more, no mutual influence takes place).

If zinc atoms are thermally excited at 5600 K instead of at 6100 K the ratio of I_{5600}/I_{6100} amounts to 0·26 according to

$$I_{5600}/I_{6100} = \{\exp. (-V_e/5600k)\}/\{\exp. (-V_e/6100k)\}$$

where $V_e = 7·75$ eV and $k = 0·855 \times 10^{-4}$ eV/°C.

This means that in Eq. (9) the intensity has to be multiplied with $1/0·26 = 3·8$. In practice a factor of 2·5 has been found which is lower than 3·8, because of the fact that zinc has been excited partly at a temperature of 6100 K followed afterwards by an excitation at a temperature of 5600 K (compare the boiling points of Zn and NaOH).

Figure 7 contains a summary of correction factors experimentally found for changes of the temperature of excitation (compare Fig. 2, Section 1.2).

It may be noted that high boiling substances, although their ionization potential can be <7 eV, will show a smaller influence on the temperature of the core than low boiling substances with the same V_i do, because of the smaller contribution which they make to the atomic population of the measuring volume.

In Section 4.2.2 correction factors related to changes of the excitation temperature are mentioned in many cases.

2.5. Matrix effects. Shifting of atom-ion equilibria (application of Saha's equation; for original literature see [20])

In Section 1.6 (Table 4), examples have been given (calculated after Saha) of the equilibrium at a given temperature:

$$M \leftrightharpoons M^+ + e$$

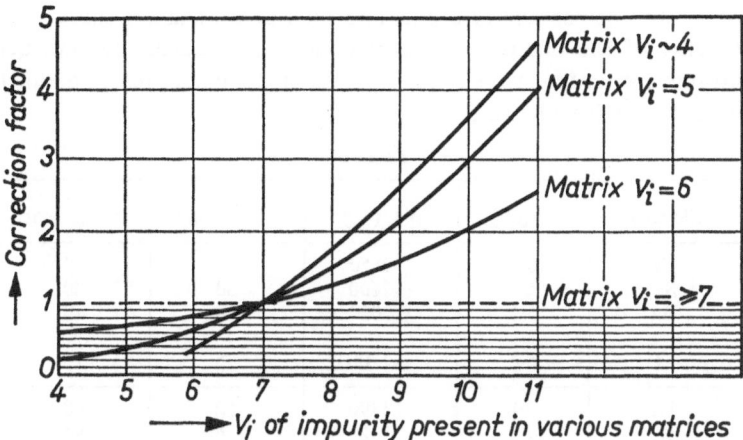

Fig. 7. Correction factors for impurities (V_i varying from 4 to 10 eV) present in matrices (V_i = 4, 5, 6 and >7 eV resp.). Concentrations of impurities found have to be multiplied with these factors.

where M means the metal atom, M^+ the metal ion and e, the electron delivered by the atom M. The lower the ionization energy of the element M, the more the equilibrium shifts to the right at a given temperature (see also footnote on p. 18). If the concentration of electrons $[e]$ remains constant, no changes in $[M]$ and $[M^+]$ take place and neither the intensity of the atom lines originating from M nor the intensity of the ion lines originating from M^+ changes.

However, if atoms are added to the measuring volume showing a low ionization energy and increasing in this way $[e]$, the atom-ion equilibrium of M is shifted to the left, which means an increase of $[M]^*$, and therefore of the intensity of lines. Elements M with a high $V_i (>7$ eV) are insensitive to added elements with a $V_i < 7$ eV, unless the concentration of the added material is so high that a significant decrease of the excitation temperature has been established.

It will be seen in Section 4.2.2 that corrections have to be made only for elements M with a low $V_i (<7$ eV). This concerns the alkalis, the alkaline earths and some other elements. It is *arc lines* of this group that have been most investigated. They originate from the mantle of the arc (Section 1.6.4). In Table 10 a summary is given

* And therefore of the intensity of atom lines and a decrease of $[M^+]$.

of V_i, the percentage of atoms and ions present in the measuring volume of the mantle (no other mineral substances are present) arranged in order of the boiling points of the oxides (earth alkalis) or of the hydroxides (alkalis).

Table 10

Element	V_i (eV)	b.p. (°C)	%-age atoms	%-age ions
Ca	6·1	2850	91	9
Sr	5·7	~2600	80	20
Ba	5·2	2000	60	40
Li	5·4	~1700	75	25
Na	5·1	1390	52	48
K	4·3	1320	6	94
Rb	4·2	700?	4	96
Cs	3·9	700?	0	100

Because of the relatively low concentration of the alkaline earths present in the test volume during their evaporation (their boiling point is high; see Section 1.4.3; Fig. 4), this group is less sensitive to additions than the alkalis. However, at the same time, the degree of ionization dependent upon the value of V_i plays a role in the same sense. The position of Li added as Li_2CO_3 (boiling point in between both groups) explains its use as a stabilizing element generally applied.

Spark lines of the alkaline earths, originating from the core, appear to be very sensitive to a complete absence of other elements. Determinations of K- and Q-factors have been carried out in matrices of the same kind (see Section 4.2.2), which means that in the atom-ion equilibrium, $[M]$ is large and $[M^+]$ is small. If now, in consequence of absence of other elements, $[e]$ becomes small then the equilibrium is shifted to the right which results in a large increase of the intensity of ion lines. This increase can be 10- to 20-fold and therefore in such particular cases as in the determination of blanks, results have to be multiplied with 0·05 or 0·1 (cf. for example Section 4.2.2 under Q-factors of spark lines of Ca). A relatively longer time of residence around the cathode is in many cases not excluded. For that reason the well-known concept of cathode layer effect ('Glimmschicht') may be mentioned here.

2.6. Matrix effects. Speed of evaporation and carrier effect; sputtering, occlusion

2.6.1. If one looks through Section 4.2.2 and summarizes number and magnitude of correction factors bearing upon change of temperature of excitation (Section 2.4), shifting of the atom-ion equilibrium (Section 2.5) and change of speed of evaporation, etc., (this section) the following results will be found (Table 11).

Table 11

Corection factors mentioned in Section 4.2.2 (taken from nominator or denominator).

	Number	Mean value
Section 2.4 (change of temperature)	10	2·0
Section 2.5 (atom-ion shifting)	15	2.6
Section 2·6 (S_i; carrier effect)	27	2·2

Two conclusions can be drawn: the mean value of the correction factors in each case above is of the same order. The significance of S_i and carrier effect is largest in view of the number of applications.

2.6.2. In 2.4 it was mentioned that mutual influence of constituents of a sample was small or negligible if boiling points differed by $\geqslant 1000$ °C. Let us take an example of the evaporation of Cu and Zn from brass. Table 12 contains some data for both elements.

Table 12

Example of carrier effect

Brass contains	V_i (eV)	b.p. (°C)	S_i (rel. cm/s)	Correction factor experimentally found*
Cu 67%	7·7	2300	80	×1·2
Zn 33%	9·4	900	350	×1

* See 4.2.2 under Cu.

The correction factor of 1·2 cannot be ascribed to a change in excitation temperature (see V_i and Fig. 2). According to its

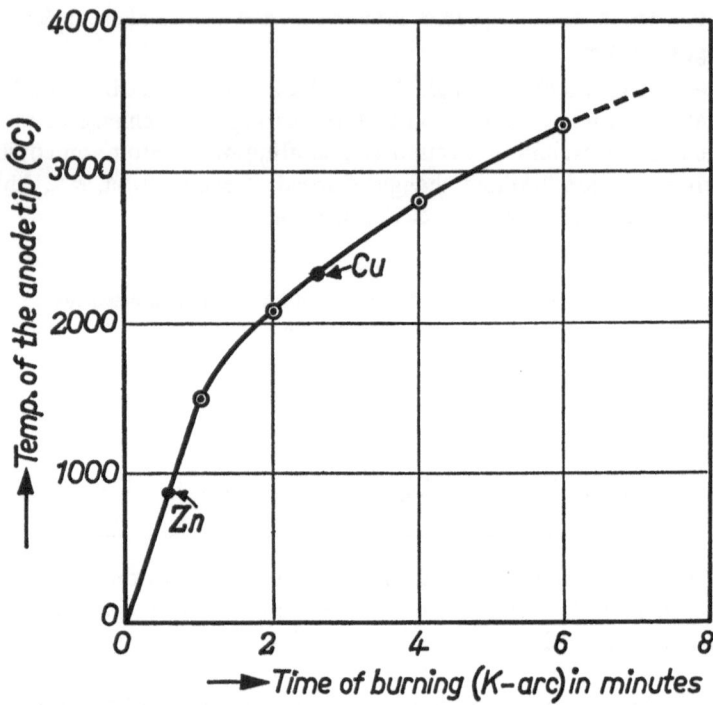

Fig. 8. The temperature of the anode tip as a function of the arcing time (*K*-arc) in minutes; the temperature has been determined by means of placing various substances with different m.p. in the crater of the *K*-anode.

boiling point Zn starts to vaporize after about a $\frac{1}{2}$ minute burn of the arc (see Fig. 8). If 5 mg Zn were present, after less than 12 seconds this quantity should have been completely 'burnt away' and if there were no mutual influence two minutes later the Cu should have also started to vaporize. In fact, however, part of this metal has been carried away during the evaporation of zinc. If the carrier effect were complete, a correction factor of 350/80 should have been applied to the content of Cu found. Because of the short period of mutual influence a factor of 1·2 has been found by experiment.

Other examples of S_i and carrier effect can be found if a relatively low boiling element such as Te is present. Then the element carried away by Te is passing the test volume in a shorter time and results of measurements also have to be multiplied by a factor >1.

The factors determining the magnitude of S_i found by experiment will now be considered.

Expressed in K Eq. (2) reads as follows:

$$\log S_i = 3{\cdot}06 - 0{\cdot}44\, T_{\mathrm{bp}(K)} \times 10^{-3} \tag{14}$$

It can be accepted that, in case of large samples, the rise of the temperature of the anode takes place in steps, each one corresponding with the boiling point of the evaporating element or compound. However, if the anode has been loaded with small samples such as 5 mg when the heat of evaporation Q_5 is also small then the temperature rises smoothly (see Fig. 8). The speed of evaporation, of the element or compound in question, depends on the position of the boiling point on the temperature curve (Fig. 8). As the temperature rise is steep at the beginning of burning of the arc, low boiling elements will evaporate with a higher speed than refractory materials where the temperature rise of the anode is even.

The mean speed of evaporation is determined by the mean pressure with which the material is evaporated. For example, 5 mg of tin evaporate in 30 seconds and the temperature gradient of the anode at the boiling point of tin (2270 °C or 2540 K) amounts to 2° per second. This value has been taken from Fig. 8, which has been constructed by measuring the time necessary to melt small amounts of various materials placed in the crater of the anode. Temperature time gradients are as follows:

Temperature of the Anode		Temperature time Gradient
°C	K	deg/s
<900	<1170	>12
900	1170	12
1500	1770	$6\tfrac{1}{2}$
1900	2170	4
2300	2570	2
2600	2870	$1\tfrac{1}{2}$
~3000	~3270	nil

In order to know the *mean pressure* with which tin evaporates we have to determine its pressure after *half* the total time of evaporation, i.e. $\tfrac{1}{2} \times 30 = 15$ seconds. During that time, the temperature has been increased from 2540 K to $2540 + 15 \times 2 = 2570$ K. Evaporation at temperatures below the boiling point has been neglected.

We now have to know the vapour pressure of tin at that temperature. After consulting [38]—at 2500 K, $p_{\mathrm{Sn}} = 10^2$ Torr; at 2960 K, $p_{\mathrm{Sn}} = 10^3$ Torr—it can be stated (Clausius-Clapeyron) that:

$$2{\cdot}3 \log \frac{p_{2960}}{p_{2500}}/(2960 - 2500) = 2{\cdot}3/460 = 0{\cdot}005$$

Now p_{2570} can be expressed into $p_{2540} = 1$ atmosphere according to

$$2{\cdot}3 \log \frac{p_{2570}}{p_{2540}}/(2570 - 2540) = 0{\cdot}005 \text{ or } p_{2570} = 1{\cdot}2 \text{ atmosphere}$$

4

Table 13

Element	T_{bpK} (K)	S_i (cm/s)	t (s)	Q (cal/mole)	Q/T_{bpK} (Trouton)	Q_5 (cal/5 mg)	Q_5/t	f (factor)	t_{corr}	Q_5/t_{corr}
Ag	2220	120	21·8	58200	26·2	2·7	0·12	1·3	28·3	0·095
Au	2870	65	21·9	84200	29·3	2·1	0·10	1·1	23·9	0·09
Co	3170	40	119	107500	34·0	9·1	0·08	1·0	119	0·075
Cr	2470	100	54	62500	25·2	6·0	0·11	1·4	75·5	0·08
Cu	2580	80	55	66900	25·8	5·3	0·10	1·4	77	0·07
Fe	3270	40	125	107000	32·7	9·6	0·075	1·0	125	0·075
Ga	2270	120	33·6	55100	24·2	4·0	0·12	1·37	46	0·085
Mn	2170	120	42·5	54300	25·0	5·0	0·12	1·63	69·2	0·07
Ni	3170	45	106	100500	31·7	8·5	0·08	1·0	106	0·08
Sn	2540	80	30	64500	25·4	2·7	0·09	1·2	36	0·075
Zn	1180	350	12·2	32000	27·1	2·4	0·20	2·3	28·3	0·085
					mean 27·2		mean 0·11			mean 0·08
					S.D. = 11%		S.D. = 35%			S.D. = 10%

$S_i = 280,000/At$ cm/s; A = atomic weight; t = time of evaporation.

Therefore, the mean pressure of tin during its evaporation amounts to 1·2 atmospheres or the speed of evaporation has been found, $f = 1·2$ times too high. This means that the time of evaporation at one atmosphere (t_{corr}) amounts to $f \times t = 1·2 \times 30 = 36$ seconds.

In Table 13, values of f and t are given, the ratio of the heat of evaporation (Q) in cal/mole and the boiling point (K) in order to confirm Trouton's rule. In the last column of the table it is shown that t_{corr} is determined by the heat of evaporation of 5 mg sample, because Q_5/t_{corr} shows a mean constant value of 0·08 (S.D. = 10%)—Values of Q_5/t differ much more (S.D. = 35%).

We now have the following numerical values at our disposal:

$$Q_5 = 5Q/1000A$$
$$Q = 27·2T_{bp(K)}$$

It follows:
$$Q_5 = 0·136T_{bp(K)}/A \tag{15}$$

Further, we know that

$$S_i = 280000/At \text{ (according to Eq. (1))}$$
$$t_{corr} = f \times t$$
$$Q_5 = 0·08t_{corr}$$

It follows:
$$Q_5 = 22400f/A \times S_i \tag{16}$$

The equations (15) and (16) can be set equal.

Therefore: $\quad 0·136T_{bp(K)}/A = 22,400f/A \times S_i$

or
$$S_i = 165f/(T_{bp(K)} \times 10^{-3}) \tag{17}$$

By putting graphically values of S_i after Eq. (14) against those of Eq. (17) we obtain a fairly good agreement between both values (see Fig. 9).

Comment. In case of element concentrations of less than 100%, say b%, the righthand parts of Eq. (15) as well as of Eq. (16) have to be multiplied by $b/100$, but the result (Eq. (17)) remains the same. Therefore S_i is independent of the concentration of the element present in the sample, provided that a fixed amount of it, in our case 5 mg, is examined.

2.6.3. *Spluttering*

The effect of spluttering can take place if, at a given temperature of anode tip, that of the sample has become so high that the volatile impurity shows a reasonable vapour pressure whereas the main component is just molten. Table 14 gives some information with regard to spluttering of Pb or Sn (oxide) present in steel.

Table 14

	Melting point (°C)	Boiling point (°C)
Pb	327	1620
Sn (oxide)	232 (1130)	2270 (1850)
Fe	1535	(3000)

Other causes of spluttering have been described in Section 1.3 (Ag, Au).

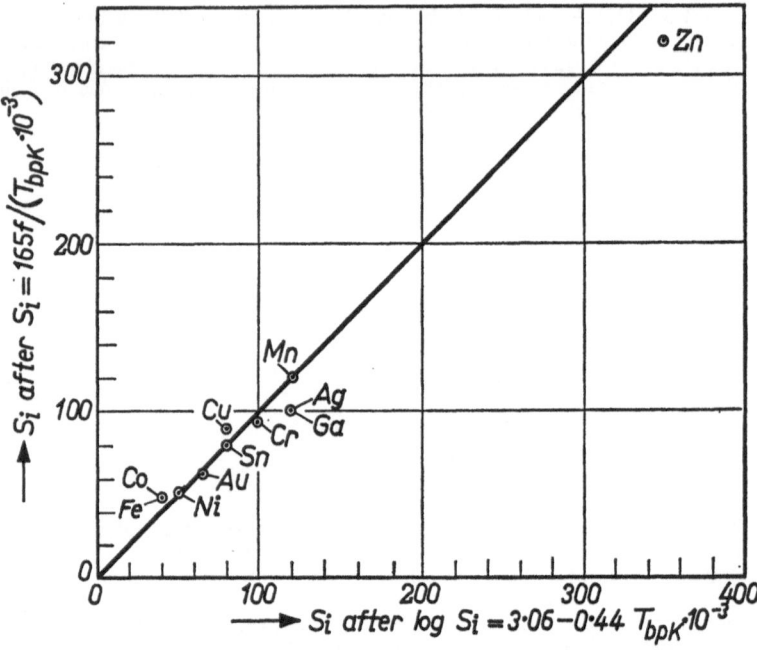

Fig. 9. Values of S_i calculated after Eq. (14) (abscissa) plotted against those calculated after Eq. (17) (ordinate) show a fairly good agreement.

2.6.4. *Occlusion* by refractory materials

Arsenic present in tungsten forms a very stable As-W-compound, which occluded by tungsten (carbide) with a b.p. of 5900 (6000) °C, does not evaporate completely. It is impossible to obtain a complete evaporation, as has been described in [22], because these authors had to deal with a refractory material which had a lower boiling point. They have chosen the addition of Ga_2O_3 (b.p. somewhat lower than 3500 °C) to determine the impurities present in uranium oxide (which is reduced in the anode to the metal itself— b.p. about 3800 °C) and, as the boiling points are close to one another, a carrier distillation will take place as described above. If the boiling points of U and the added substance differed largely, no effect should have been found. (A modified method is the addition of halogen compounds, then a chemical reaction takes place. The impurities are brought in the halide form and evaporate easily as such).

Note. Ga_2O_3 will also have been chosen because of the sparse spectrum of Ga; the number of its lines between 2000 Å and 8000 Å (L2–8000) amounts to 135.

Chapter 3

Spectral Lines and Measurements; Instruments and Techniques

In the early days of spectrochemical analysis, no adequate tables of wavelengths were available and the identification of lines to measure was a major problem. Now tables are adequate, this is reduced to the choice of lines to measure, and the method of measurement chosen is an important decision. The s.p.d. method calls for the selection of lines of moderate or low intensity, and avoids the lines of high 'sensitivity' which are used for trace analysis only, since these are often liable to self-absorption. Choice of lines and the principles of the choice, are discussed in this chapter. Section 3.1 deals with the selection of lines for each element and in Section 3.2, the reader can see the s.p.d. scale. This looks like an arbitrary scale of tints of grey, but users will soon find that it is extremely hard to find a faint spectrum line which does not match one step or another of this scale, since the steps were selected for this use. The practical equality of the steps was verified after a long series of preliminary practical tests (see [26]) and the practical advantage of using a scale which corrects the line considering its background is a vital one in increasing the precision and speed of the analysis. The scale was intended for use with a projector which throws the spectrum down onto a horizontal surface. The uniformity of the projector illumination, its cleanness, and of the room light level all need careful attention.

Since general principles are hard to understand without examples, an example of steel analysis is worked out in Section 3.3 and this should be followed through to make sure that it is understood before the rest of the book is used. Section 3.4 should be read to realize how much the type of spectrograph employed contributes to the accuracy of the results which this method will give. In Section 3.5 it is not the kind of spectrograph used but the way of focusing the radiation into it and also the way the sample is burnt that are changed to lower the limit of detection for special analytical problems, where precision is more important than speed. By making a fresh set of six spectra per sample and applying the s.p.s. factors (listed in Table 9) the scope of quantitative analysis can be extended downwards by a factor of at least 20, and the limit of detection by about 50.

3.1. Choice of spectral lines

3.1.1. According to Einstein's law (eV $= h\nu$), a wavelength of light emitted by an excited atom is determined by the ratio 12380/$(V_{eh} - V_{el})$, where V_{eh} is the excitation potential of the high level and V_{el} is the excitation potential of the low level of the excited state of the atom (see the Appendix for details). If a transition takes place between an excited state and the ground state of the atom, V_{el} is zero.

Spectral lines of elements situated in the left part of the periodic system of elements, which show a low value of V_i, can be found, according to the formula just mentioned in regions of longer wavelengths. Similarly the spectral lines of elements situated in the right part of the periodic table which show a high value of V_i (also V_e) can be found in regions of shorter wavelengths.

As the second ionization potential of all elements is higher than the first, excitation energies of singly charged ions can also reach a higher value than V_e of atoms, but emission is limited to 9·5–10 eV. Generally ion emission will be shifted somewhat to shorter wavelengths; the alkalis show an extreme example.

The wavelength range 2500–3500 Å (mostly used) offers the possibility of measuring lines, whose transition energies $(V_{eh} - V_{el})$ lie between 12380/2500 $= 5$ and 12380/3500 $= 3\cdot5$ eV. Many elements show transitions between these values. Lines of the alkalis are situated at longer wavelengths and those of the transition elements are found more at shorter wavelengths.

3.1.2. In order to avoid disturbing influences caused by line-broadening or self-absorption it is preferable to measure *weak* lines with a density up to 0·4–0·6 (see Fig. 10, *H* and *D* curve). In cases of sparse spectra it is sometimes necessary to measure higher densities.

In general it is advised to measure as many lines as possible within these limits.

3.1.3. As has been pointed out in Section 1.6.4, measurements of lines originating from the core are more accurate. For elements with a V_i round 7 eV (Section 1.6.2; Fig. 5) arc- as well as spark-lines have to be measured and results have to be averaged. An arc-current deviating from the normal one (10 A) can cause a significant difference between mean results of measurements of arc- and of spark-lines. This can also be observed in the determination of the E.C.F. (Section 3.3), which is found by measurement of three Fe I lines (2645·42, 2667·91 and 2815·50 Å) and one Fe II line (2883·70 Å). All four lines have to give the same result ($C = K \times I$— Section 2.2, Eq. (9)). Significant differences in arc- and spark-line measurement are caused by an incorrect arc current.

3.1.4. Spectral lines generally suitable for the concentration deter-mination have been assembled in the Appendix with some remarks concerning coincidences of lines. However, it is impossible to mention them all. One should consult also [23, 24].

In order to find spectral lines easily, it is advised to indicate them permanently in photographed iron spectra, for instance those taken from the well-known atlas of Junkes and Gatterer (iron spectra are always present on the photographic plate or film under examination for the determination of the E.C.F.).

3.2. Determination with the s.p.d. scale, with the microphotometer and with photomultipliers

In order to determine the 'integrated intensity' (integrated over the period of complete evaporation of the element in question) of a spectral line it is necessary to carry out two measurements. The determination of the maximum value of maximum D_{l+b} (density of line plus background) and the determination of D_b. These values are transferred into I_{l+b} and I_b via the H and D curve (Fig. 10). By subtraction $I_l = I_{l+b} - I_b$.

This calculation normally used can be simplified by applying the s.p.d. scale (Fig. 11 [25, 26]), which automatically corrects for back-ground densities up to 0·10. Moreover, as measurement of weak lines is preferable (see Section 3.1.2), this visual method of deter-mining I_l suits ideally, because the human eye is capable of distin-guishing small density differences very accurately in the range up to $D = 0·6$ (a c.o.v. of 7%). A further advantage of this method of inspecting spectra is the resulting combination of qualitative and quantitative examination.

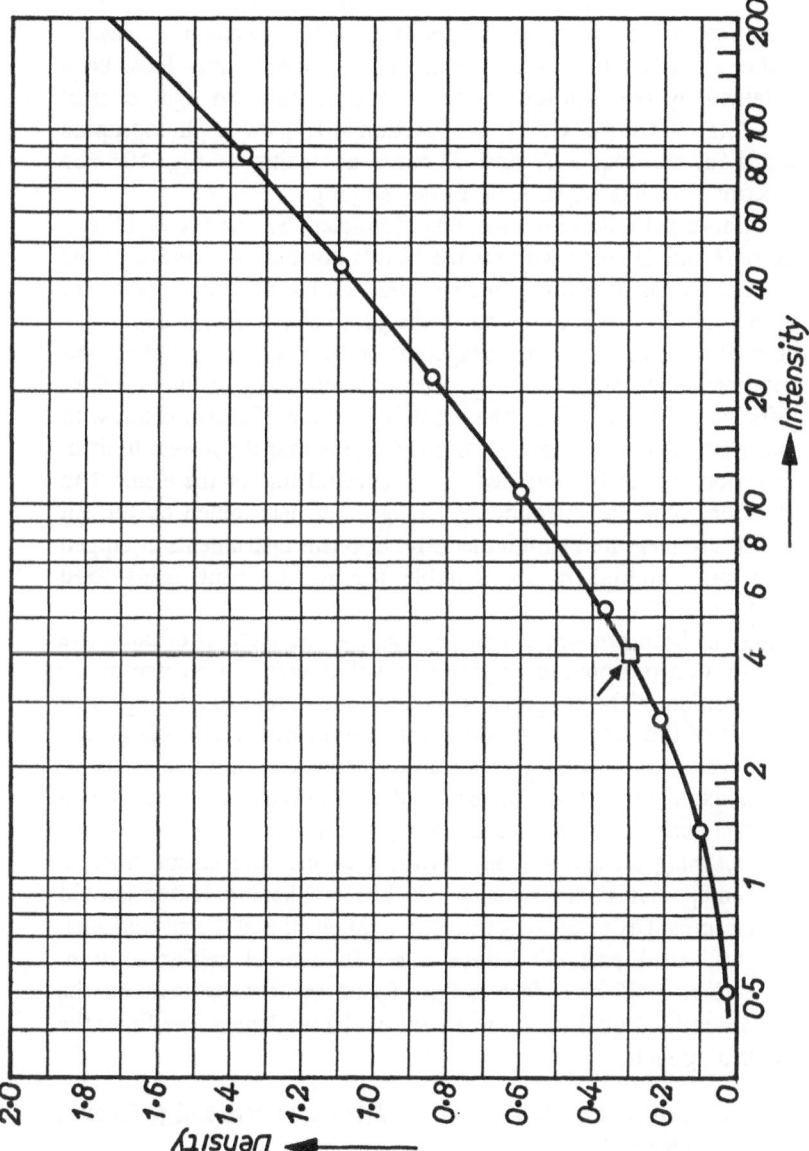

Fig. 10. *H* and *D* curve of the photographic emulsion, developed in such a way that a density of 0·3 corresponds with an intensity of 4 and that further the contrast of the emulsion gamma amounts to about 0·9.

The s.p.d. scale, shown in Fig. 11, has been verified by comparing them with a hundred spectral lines measured by means of a densitometer according to $I_l = I_{l+b} - I_b$. These spectral lines have been obtained by the exposure of 'a 5 mg iron standard light source' (K; 13.1; 7 Cm; 7 C; 7) (see Section 4.1.7) on a photographic emulsion, showing a H and D curve according to Fig. 10. For spectral lines like s.p.d. scale lines, see [26].

If photomultipliers are used, (the characteristic curve resembling a steep H and D curve without the 'foot') not only the centre of the spectral line is measured, but also the total surface of it (slits are wider). The advantage consists in the possibility of measuring broad lines. It is, however, disadvantageous not to be able to measure lines lying near each other (visually and photometrically distinguishable; principle of Rayleigh) as the resolving power of instruments with wider slits is poor. Another disadvantage is that the lowest limit of detection cannot be obtained as the spectral line of the element in question, plus the surrounding background determined by the slit width, are measured simultaneously. Spectral instruments equipped with photomultipliers are suitable for measurement, using fixed programmes.

As has been pointed out in Section 1.4, the method of complete evaporation (no pre-arcing period) involves constant change in the composition of the arc gases. Therefore, measurements during short periods of time are inadmissible. Integration over the whole period of 'burning' is necessary and photographic emulsions lend themselves excellently. It is not excluded that automation of universal spectral analysis can be effected along this way.

Weakening sector. If it is inevitable to measure strong lines, a weakening sector can be applied. The weakening factor should be determined by measuring the ratio of intensities of the unweakened and weakened part of various lines. In case of self-absorption, corrections of K- or Q-factors have to be made dependent on the true intensity (equal to the intensity of the weakened line times the weakening factor).

3.3. Calculation of analysis results applying K-factors and Q-factors; limits of detectability

3.3.1. K-method

It is asked to analyse a stainless steel for Cr, Ni and Fe. Spectra are taken (in duplicate) according to the prescription mentioned at the end of the iron K-, Q-table; Appendix: K, 13.1; 7 C m; 7 C; 7; see Section 4.1.8, which means: K-method (5 mg filings of the steel sample to be analysed); anode shape (Fig. 13.1); mix well with about 7 mg graphite (C) powder in the crater of the anode with a small spatula; cover this mixture with about 7 mg graphite (C) powder, and 'burn' for 7 minutes (arc current 10 amps).

In both spectra, thus obtained, low-intensity lines (see Section 3.1.2) of Fe, Cr and Ni have to be measured and, as chromium arc lines mainly originate from the mantle of the arc ($V_i = 6.74$ eV; see further 1.6.4 and Fig. 5) Cr I and Cr II lines also have to be measured. Results of these measurements as well as the determination of the concentration of these elements are shown in Table 15. The elements Si, Mn, Mo, Cu, Co and V have been found in exactly the same way. The sum of all concentration values mentioned in column 4 amounts to 132·8%. The main cause of this high value is because of the photographic plate or its development. The sum of 132·8% has to be brought back to 100% (as all elements present have been measured) and therefore an Internal Correction Factor (I.C.F. $= 100/132\cdot8$) has to be applied.

If only one element had been measured, the External Correction Factor (E.C.F.) has to be determined and this is done by taking spectra of pure iron in duplicate and measuring the same four Fe-lines mentioned in Table 15. The E.C.F. and I.C.F. can only differ because of the weights of a 5 mg sample and 5 mg pure iron. All plates have to be calibrated in this way, but if a complete analysis is carried out the I.C.F. is preferable. In 3.3.2 (analysis by means of Q-factors) the E.C.F. has always to be applied, because adding up to 100% is impossible.

It is worth-while to note that by adding up to 100% the highest concentrations found are the most accurate. Suppose that for the impurities a sum of 2·7% had been found in lieu of 1·35% (Table 15, column 4), which means a relative difference of 100%. The sum mentioned in column 4 should have been 134·1 and the I.C.F. should have been $100/134\cdot1 = 0\cdot745$ instead of $100/132\cdot8 = 0\cdot752$, a difference of 0·9%. Adding up to 100%, increases the accuracy of the highest concentrations determined.

If a sample containing oxides (or salts) for instance had been analysed, the elemental concentrations mentioned in Table 15,

Table 15

The Determination of Fe, Cr and Ni in Stainless Steel

(1) Element λ(Å)	(2) Intensity		(3) K-factor (%)	(4) Concentration (%) mean	(5) Concentration (%) × I.C.F.	(6) Chemical result (%)
	Duplicate	Mean				
Fe 2645 I	3·93 4·27	4·10	25·1	103		
2667 I	3·43 3·50	3·47	27·1	94 ⎫		
2883 I, II	3·30 3·10	3·20	28·6	91·5 ⎬ 96·3	72·5	71·6
2815 I	3·30 3·23	3·27	29·6	96·7 ⎭		
Cr 2785 II	2·06 1·90	1·98	9·0	17·8		
2700 I	2·60 2·44	2·52	9·0	22·6$_5$ ⎫		
2642 I	2·88 2·53	2·71	7·8	21·1$_5$ ⎬ 23·3$_4$	17·6	18·3
2913 I	4·73 4·08	4·41	6·4	28·2		
2935 II	5·45 4·70	5·08	5·3	26·9 ⎭		
Ni 2540 I	1·99 1·92	1·96	6·0	11·7$_6$ ⎫		
2802 I	1·96 2·13	2·05	5·5	11·3 ⎬ 11·7$_8$	8·9	8·98
3145 I	10·23 10·27	10·25	1·2	12·3 ⎭		
Si, Mn, Mo, Cu, Co and V				1·35	1·0	0·97
				sum 132·8	→100	

For K-factors see Tables, Section 4.1.1.

The concentration found by measuring each line in duplicate has been calculated according to Appendix (9), Section 2.2. No extra corrections had to be made (see Section 4.2.2 under Fe, Cr and Ni).

The final result (concentration found by multiplication with I.C.F. is shown in column 5; results chemically found are shown in column 6 for comparison.

Fig. 11. The s.p.d. scale (Standard Paper Density scale): a number of lines of increasing density and of the same shape as spectral lines [26]. The line of the s.p.d. scale is illuminated by the light source of a Hilger projector filtered by the background adjacent to the spectral line to be measured. In this way automatic correction for background (within density limits) is achieved.

column 4 would have to be converted into those of the oxides (or salts) (see Section 4.5) and their sum brought to 100%, as has been shown for the elements in column 5. For the limits of detectability (K-method), see Section 3.3.4.

3.3.2. *Example of a calculation* of the concentration if K- or Q-factors gradually change (resonance lines, Section 2.3).

The spectral line 3273 of Cu (present in CdS; see Appendix under Cu) shows an intensity of 10; the E.C.F. shows a value for the photographic emulsion of 0·8. This means that the real intensity amounts to 0·8 × 10 = 8. The K-factor belonging to an intensity of 8 amounts to 0·0036 and the final concentration amounts to 8 × 0·0036 = 0·029% Cu.

3.3.3. *Q-method.* Concentrations are calculated in the same way as has been shown in Section 3.3.1 with the exception that only the E.C.F. has to be applied. Generally a few lines are available for measurements and results are less accurate.

3.3.4. *Limits of detectability.* If the most sensitive lines are absent, it is possible to give the limit of detection by multiplying Q-factors by 0·3. In a clear spectrum, a line showing an intensity of 0·3 ($D = 0·03$; see H and D curve Fig. 10) is just detectable both visually and photometrically; if the background density amounts to about 0·1 (wavelength range 2480–3100 Å); for higher backgrounds wavelength range 3100–3400 Å) a factor of $\frac{1}{2}$–1 has to be applied. For final results multiply by the E.C.F.

Further discussions of limits of detectability are given in Sections 3.5 and 4.5.

3.3.5. If *spectrographs other than* prism instruments are used, conversion factors also have to be taken into account.

3.4. Comparison of various spectrographic instruments; conversion factors

K- and Q-factors given in the Appendix have been determined by applying a large quartz Hilger spectrograph (E492 or E489). If other instruments are used it is necessary to know the sensitivity of wavelengths through the region 2500–3500 Å normally used. To this aim an 'Fe standard light source' has been applied, the same as has been used for the determination of the E.C.F. (see Section 3.3.1). Various light source slits were chosen in such a way that the four iron lines applied for the determination of the E.C.F. showed densities equal to about 0·3 ($I = 4$; see Fig. 10).

Table 16 shows the results of measurements of intensities of iron lines originating from the 'standard light source' expressed in (relative) units derived from Fig. 10. An example is shown also of the conversion factors to be applied to measurements obtained with a grating spectrograph (A.R.L. 2nd order). As has been shown in [27] prism instruments show a conversion factor equal to unity.

Conversion factors $I_{\text{Hilger}}/I_{\text{ARL}}$ are given in Table 16, column 4.

The concentration found, according to Eq. (9) (see Section 2.2), has to be multiplied by the conversion factor for the wavelength in question (see Fig. 10, where a curve is shown drawn through various measuring points taken from Table 16, column (4)).

For analysis results found with various spectrographs and worked out according to the method above-mentioned, see [27].

3.5. Further decrease of the limits of detectability by changing optics; the s.p.s. method

3.5.1. In Section 3.3.4, it was mentioned that the density of a spectral line of 0·03 was just visually as well as photometrically detectable in a background of 0·1 or less (see H and D curve, Fig. 10).

Now the intensity, I_l, of a line (determined by measuring densities and converting them into (relative) intensities via the H and D curve) is found by subtracting I_b (intensity of the background near the line) from I_{l+b} (intensity of line plus background): $I_l = I_{l+b} - I_b$ (see also Section 3.2).

Below the limit of detection, the difference between I_{l+b} and I_b is so small that density measurements show no significant difference. However, the difference ($I_{l+b} - I_b$) can be increased with multiplication by m, according to: $mI_l = mI_{l+b} - mI_b$. Density measurements can then be carried out on the straight part of the H and D curve.

A high value of this multiplication factor, m, can be attained in three ways [11]:

1. by taking 25 mg sample instead of 10 mg; m has been found to be 2.

Table 16

Conversion Factors of a Grating Instrument (A.R.L. 2nd order)

(1) Fe I *lines* *in* Å	(2) *Transition* *in* eV		(3) *Intensity* *(rel. units)* I_{Hilger}	(4) I_{Hilger}/I_{ARL}
2493·99	⎰ 5·96 ⎱ 5·90	1·01 ⎱ 0·95 ⎰	12·9	
2495·86	5·80	0·86	12·7	
2516·56	5·86	0·95	4·1	
2560·55	5·83	1·01	2·9	
2579·26	⎰ 5·70 ⎱ 5·77	0·91 ⎱ 0·99 ⎰	2·2	
2612·77	4·77	0·05	7·3	0·98
2636·47	5·59	0·91	5·6	0·98
2667·91	4·71	0·09	3·7	1·00
2710·54	6·15	1·60	10·8	1·00
2781·83	5·42	0·99	15·5	0·97
2815·50	5·98	1·60	3·2⁵	1·04
2817·50	5·33	0·95	9·0	0·98
2886·31	5·83	1·55	2·8	1·00
2988·46	5·61	1·48	2·6	1·00
3014·17	5·05	0·95	2·4	1·00
3029·23	5·62	1·55	3·3⁵	1·08
3078·43	6·48	2·47	5·1	
3144·48	6·38	2·46	6·0	
3153·20	6·35	2·44	12·6	
3161·94	6·29	2·39	14·2	1·18
3165·00	6·31	2·41	4·0	1·19
3207·09	6·24	2·39	2·9	1·11
3276·47	5·96	2·19	3·7	1·32
3284·58	5·95	2·19	10·1	1·21
3298·13	5·96	2·21	15·1	1·20
3324·54	6·11	2·39	5·4	
3325·46	6·15	2·44	3·5	1·25
3356·40	5·95	2·27	4·6	
3396·97	4·59	0·95	11·9	
3401·52	4·54	0·91	29·6	
3415·53	5·83	2·21	10·7	

2. by focusing the arc on the slit of the spectrograph instead of on the collimator lens. In the case of the application of the Hilger large quartz spectrograph (E492; a prism instrument) the increase appeared to be thirteen times.

3. by superimposing a number of spectra on one another, limited by densities of <2 to be measured on the H and D curve, the maximum superimpositions appeared to be 4. (For further possibilities of increasing m in case of photocells, see [31].)

In this way m can amount to $2 \times 13 \times 4 = \sim 100$. The method of analysis will be indicated by s.p.s.: *s*uperimposed spectra, light source *p*rojected on *s*lit of spectrograph. It does not mean, however, that a decrease of the limit of detection by a factor of about 100 will be found. If a significant value is to be given, it is necessary to consider the variability of the background and that of the line. The presence of a line can be significantly determined (95% confidence level) if the difference with m-fold amplication complies with:

$$mI_{l+b} - mI_b > 2\sigma_b \sqrt{2} \tag{13}$$

see also [32].

It may be expected that at the limit of detection, σ_b and σ_{l+b} do not differ very much and hence σ_b determines, according to the equation just mentioned, the limit of detection of the element of which its spectral line is situated in that particular background.

According to our experience with the Hilger instrument (and the skill of the operator, focusing on the slit of the spectrograph is somewhat difficult) σ_b amounts to about 1%; i.e. $0.01\,I_b$.

As spectra are taken in six-fold (three on one plate and three on another),* σ_b amounts to $0.01\,I_b/\sqrt{6}$ and therefore $2\sigma_b\,\sqrt{2}$ (Eq. (18)) is equal to $2 \times 0.01\,I_b\,\sqrt{2}/\sqrt{6} = 0.0116\,I_b$.

In comparison with the limit of detection by means of the Q-method (Section 3.3.4), the gain factor g_f for the region 2500–3100 Å amounts to:

$$g_f = \frac{1/3\,Q}{0.0116\,I_b\,Q/100} = \frac{2870}{I_b} \tag{19}$$

($4 \times$ superimposition is possible; I_b reaches values of 80). For the

* On two plates: in order to promote independence of measuring results.

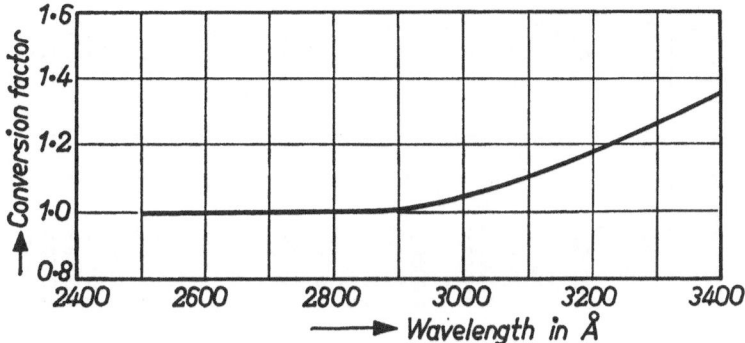

Fig. 12. Conversion factors (ordinate)—I Hilger E 492/I 1·5 metre A.R.L. 2nd order—plotted against wavelengths (in Å).

region 3100–3400 Å (no superimposition; the limit of detection is equal to $1 \times Q$) g_f amounts to:

$$g_f = \frac{Q}{0 \cdot 0116 \, I_b \, Q/25} = \frac{2150}{I_b} \qquad (20)$$

The mean value of I_b (found in the region 2500–2900 Å) in $4 \times$ superimposed spectra amounts to 60 and g_f, amounts to 2870/60 (14) = 48, nearly half the amplification of 100.

The limits of detection of the s.p.s. method for a number of element lines have been calculated according to the Eq. (19) and (20). (See Table 17.)

In practice σ_b amounts to values $\geqslant 0 \cdot 01 \, I_b$. Interferences with molecular bands or impurities, present in blanks of the graphite used, can cause an increase of σ_b up to 5 times its original value and a decrease of g_f with a factor of 5.

3.5.2. Some examples of the application of the s.p.s. method.

(a) The determination of Ga present in CdS. A hundredfold amplification was obtained as described in Section 3.5.1 (analyses in 6-fold). Q (Ga 2943) = $0 \cdot 0002\%$.

(b) The determination of Sb present in Ge. Pure Ge and a preparation containing $0 \cdot 00075\%$ Sb enabled us to prepare (Fig. 13.2) mixtures containing $0 \cdot 00065$, $0 \cdot 00015$ and $0 \cdot 00008\%$ Sb in the

Table 17

Limits of Detection; s.p.s. Method

Element	Wavelength in Å	Q-factor in %	I_b	Eq.	g_t	Limits of detection in %	(s.p.s. method) in ppm
Ag	3280	0·000015	65	(20)	33*	0·000005	0·005
As	2780	0·015	60	(19)	48	0·0001	1·0
Be	3130	0·00003	25	(20)	86*	0·0000004	0·004
Bi	3067	0·0003	75	(19)	38	0·000003	0·03
Co	2521	0·001	60	(19)	48	0·000006	0·06
Cr	2835	0·002	60	(19)	48	0·000015	0·15
Ga	2943	0·0003	60	(19)	48	0·000002	0·02
Ge	2651	0·0003	60	(19)	48	0·000002	0·02
Hg	2536	0·0007	60	(19)	48	0·000005	0·05
In	3256	0·0006	50	(20)	43*	0·000015	0·15
Ir	2849	0·0025	60	(19)	48	0·00002	0·2
Mn	2576	0·0002	60	(19)	48	0·0000015	0·015
Mo	3132	0·001	25	(20)	86*	0·00001	0·1
Ni	3414	0·0002	190	(20)	11*	0·00002	0·2
Ni	3002	0·0008	65	(19)	44	0·00006	0·06
Os	2909	0·0025	60	(19)	48	0·0002	0·2
P	2535	0·01	60	(19)	48	0·00007	0·7
Pb	2833	0·0003	60	(19)	48	0·000002	0·02
Pt	3064	0·0004	75	(19)	38	0·0000035	0·035
Rh	3434	0·001	220	(20)	10*	0·0001	1·0
Rh	2555	0·015	60	(19)	48	0·0001	1·0
Sb	2598	0·002	60	(19)	48	0·000015	0·15
Sn	2839	0·001	60	(19)	48	0·000007	0·07
Ti	3241	0·002	50	(20)	43*	0·00005	0·5
Tl	2767	0·007	60	(19)	48	0·00005	0·5
V	3185	0·002	30	(20)	72*	0·00003	0·3
Zn	3075	0·02	78	(19)	37	0·0002	2·0
Zr	3391	0·001	170	(20)	12½*	0·00008	0·8

* g_t has reference to Q (not to 1/3 Q).

Table 18

Results of Analyses According to the s.p.s. Method

Preparation	Intensity of the Amplified Line	$c = Q \times I/amplification$	Wet-chemical or Physical Analysis
a1 (Ga in CdS)	8·7	8·7 × 0·0002/100 = 0·000017%	0·00002%
a2	3·1	3·1 × 0·0002/100 = 0·000006%	0·000005%
b1 (Sb in Ge)	13·9	13·9 × 0·0014/25 = 0·00078%	0·00075%
b2	12·4	12·4 × 0·0014/25 = 0·00070%	0·00065%
b3	3·4	3·4 × 0·0014/25 = 0·00019%	0·00015%
b4	1·9	1·9 × 0·0014/25 = 0·00011%	0·00008%
c1 (Fe in ZnS)	(14·4 − 12·6 =) 1·8	1·8 × 0·0006/100 = 0·000011%	0·000013%

Remark to the analysis c1: As σ_b appeared to be about twice as large as 1% of $I = 60$ (i.e. 1·1 instead of 0·6) the limit of detection (Eq. (19)) now amounts to $\frac{1}{3}Q/gf = \frac{1}{3} \times 0·0006/(2870/2 \times 60) = 0·000008\%$ Fe; the value 0·000011% found shows that Fe is significantly present.

crater of the Q-anode. Amplification (without superimposition) 25×; analyses in 6-fold. Q (Sb 2598) = 0·0014%.

(c) The determination of iron present in ZnS by means of 1 ppm Fe containing graphite electrodes. Blanks were determined by applying iron-free GeO_2 (see Section 4.5). Amplification 100×; analyses in 6-fold. Q (Fe 3020) = 0·0006%.

For results, see Table 18.

Chapter 4

Working Instructions; Reproducibility and Accuracy

This section of the book is a compilation of working instructions and tabular information. Section 4.1 begins with the selection of electrodes, and their manipulations, and the supervision of the arc during burning. Exposures of up to 8 minutes need care to maintain the arc length and position uniform throughout the exposure period. The difference between the K-method and the Q-method is fully explained here, as is sample preparation and the choice of graphite for electrodes for special materials. It is necessary to read Section 4.1 which explains the text before attempting to use the cards, of which there are enough to provide convenient lists of lines for each of the 63 elements to which the K- or Q-factor method has been applied (Section 4.2).

When the lines have been selected and measured, the accuracy of the results may be queried, and a brief note on statistical evaluation of reproducibility based on short runs of replicate tests is provided in Section 4.3. The use of the word accuracy should be confined to the agreement of an estimate with the real true value (arrived at by other methods) while reproducibility is much easier to measure. (Sensitivity is a word which is changing its usage, it was formerly used to mean sensitive to low concentrations, but now, under the influence of users of flame photometry and other chemical methods, spectrochemists are accepting the wider definition of sensitivity as the gradient of the slope of the graph of line intensity against concentration, on logarithmic scales.)

Section 4.4 has a list of conversion factors for calculating the quantities of oxides or salts from the weight percentages of elements found in analysis. Section 4.5 discusses another aspect of the 'matrix effect', the fact that the residual impurities in so-called pure graphite can only be estimated with the aid of a carrier substance such as GeO_2, and there is an estimation of 'carbon impurities' present in a sample, worked out as an example.

4.1. Practice of 'burning' of the arc; shapes of electrodes; loading of the anode; shape of electrode holders; exposure time; what to do if spluttering occurs

4.1.1. *K-arc*

It has been pointed out in the Introduction that in order to establish complete evaporation, anodes with a smaller diameter than 10 mm are needed for substances with a very high boiling point. For this reason most anodes are shaped as in Fig. 13.1 or 13.2, but for very high boiling materials the anode shape 13.4 (Harvey anode) is necessary, as it gives more reliable results. In all cases the cathode has the shape shown in Fig. 13.3 the tip of it has been undercut in order to prevent climbing of the arc. By undercutting, the tip reaches a higher temperature than without it and condensation is prevented.

The anode 13.4 is applied in case of spluttering substances (Ag, Au for example; see Section 1.3) and also to high boiling materials as mentioned before.

Advised maximum exposure times, depending on the boiling point of the main component, are given in Table 19. They are longer than t (in seconds) mentioned in Table 3, but the time necessary for reaching the boiling point of the element in question is added (see Fig. 8).

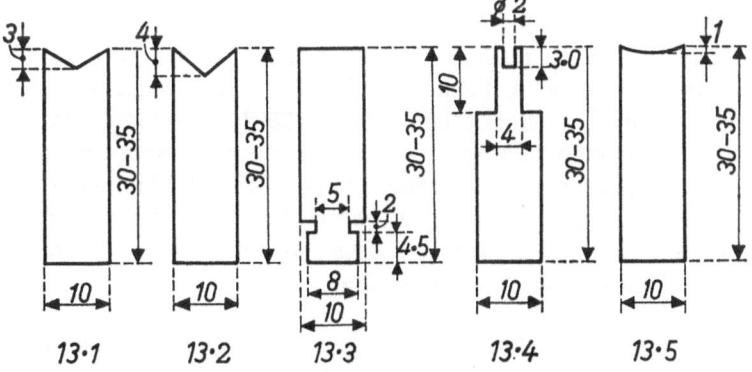

Fig. 13. Various shapes of electrodes (dimensions have been given in mm). *K*-anodes: 1, 2 and 4; *Q*-anodes: 5, 1, 2 and 4; *K*- and *Q*-cathode: 3. The anode 13·2 is used instead of 13·1, if bulky powder has to be analysed. In many cases rods of the purest graphite have a diameter of 8 mm instead of 10 mm; they can also be used.

Moreover, some molten material could be sucked into the anode and for this reason exposure times had to be taken as somewhat longer (experimentally tested).

Table 19

Maximum Exposure Time of the *K*-arc in Dependence on the Boiling Point of the (expected) Main Component*

Boiling point °C	Exposure time (minutes)	Anode shape
a. <1800	6	13·1; 13·2
b. 2100	6½	13·1; 13·2
c. 3000	7	13·1; 13·2
d. ~4000	8	13·1; 13·2
e. up to 6000	4	13·4

* For unknown substances the longest exposure time is advised.

Remarks
a–d. 5 mg of the sample is mixed with 7 mg graphite (C) in the anode.
a–b. This mixture is covered with 30 mg graphite (C).
c–d. This mixture is covered with 7 mg graphite (C).
d. Independent of their boiling points also the oxides of the rare earths; because of their low ionization potentials (the arc sticks immediately to these substances) the anode shape 13.4 causes explosive evaporation.

During burning readjustment of the electrode gap to maintain 9 mm is necessary. A diaphragm of about ½ mm ensures that only light around the centre of the gap (at the height of 4½ mm) is sent to the spectrograph (see Fig. 14). Readjustment of both electrodes into the optical axis is also necessary when the arc is burning.

The light source, diaphragm C, ensures that light originating from the cathode- and anode-layers does not enter the spectrograph; all the light is supplied from the *Q*-arc.

Q-arc (see Fig. 14 but without light source diaphragm; electrode gap 2 mm; electrodes at the same distance from the spectograph slit; *pro memoria:* 10 mg sample). In Table 20, details are given for the *Q*-arc concerning exposure time, etc. All the light of the *Q*-arc is sent to the spectrograph. Readjustment of the arc gap of 2 mm is necessary as well as control of alignment in the optical axis.

Table 20

Exposure Time of the Q-arc in Dependence on the Boiling Point of the (expected) Main Component

Boiling point °C	Exposure time (minutes)	Anode shape
a. <1800	1	13·1; 13·2
b. 1800–~3500	1	13·5
c. 3500–6000	$3 \times 1^*$	13·4

* Exposure time 3 minutes in 3 separate steps of 1 minute in order to prevent heavy background; the arc burns continuously.

Remarks
    a. The anodes 13·1 and 13·2 are also used in case of possibly spluttering coarse powders.
    c. The oxides of the rare earths are arced to completion, but the anode 13·1 is used (see Table 19, remark d).
    a–c. Graphite powder is not applied.

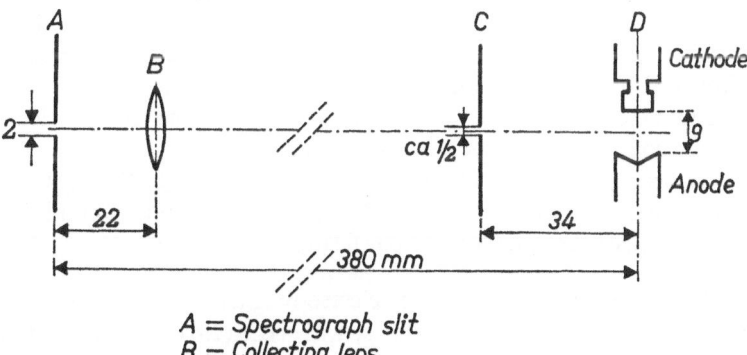

A = Spectrograph slit
B = Collecting lens
C = Diaphragm
D = Electrodes

Fig. 14. Optical lay-out (arc-diaphragm-collecting lens-spectrograph slit); when a Littrow spectrograph is used, it is recommended to set the diaphragm ½–1 mm above or beneath the optical axis. The reflections from the optical surfaces inside the spectrograph are thus diverted sideways and to a large extent absorbed by the blackened interior of the instrument.
    The diaphragm C (light source slit) has the following dimensions: length 15 mm and width about 0·5 mm. Its position is such that the light originating from a 'slice' (diameter 15 mm; thickness 0·5 mm) of the arc at half the arc gap of 9 mm is let through.

As the Q-arc is only used for the determination of low concentrations ($\sim$0·01 – $\sim$0·0001%) purest graphite rods for preparing electrodes are applied and their diameter generally amounts to 8 mm in lieu of 10 mm (K-arc).

4.1.2. *Preparation of samples and anodes*

Liquid substances like mercury are converted into sulphide or selenide by means of the addition of spectroscopically pure S or Se. In this way 5 or 10 mg are weighable. A material like gallium can best be cooled in liquid air, packed in filter paper and broken to pieces by hammering.

Materials containing volatile substances like water are freed from them by pre-heating in a furnace (120 °C, constant weight).

When blanks of graphite rods or minute quantities of impurities present in refractory materials (for example SiC and the like) have to be determined, addition of 10 mg GeO_2 (extremely pure), well mixed with 10 mg of the sample in the anode, is advised.

The maximum tolerable grain size of powders or filings amounts to 0·2 mm. Special care must be taken in the preparations of metal filings. The file must be cleaned by filing the metal in question followed by cleaning with a brush, first filings are not used. By this method, the surfaces of the metal as well as of the file are clean. The second filings can be used for the analysis.

Sometimes particles of the file fall into the filings (for instance iron particles of the file into a tough substance like aluminium). In such cases two methods of purification can be tried:

1. the application of a strong electro-magnet;
2. afterwards boiling the filings, with HCl (2M) for one minute. The acid is removed by treatment with distilled water and drying with alcohol or acetone.

In order to prevent mortar particles mixing with the samples to be powdered, it is advised to have various mortars at one's disposal. Agate and silicon carbide mortars mainly introduce Si into the sample, boron carbide mainly B, molten aluminium oxide mainly Al and other impurities [28]; in many cases small crevices present in the mortar contain material powdered before. The agate mortar used for powdering $AgNO_3$ can only be cleaned by standing with ammonia overnight. Regular control and renewed grinding are necessary.

Generally powders are not pressed into the anode. In the case of fluffy powders, a light pressure has to be applied, sometimes the volume of 10 mg (for instance, silica prepared from ethylsilicate) is so large that one needs to analyse only half the mass. The results then have to be multiplied by 2.

Addition of a known (micro) quantity of an element to a pure matrix (preparation of a 'home-made' standard) or to an impure sample (to check results by applying the 'addition method'). Because of the great possibility of loss of the additive (dissolved in water) by its adherence to the wall of the container, the wettability of the powdered pure matrix or the sample should be determined in the following way: place 500 mg (or less) in a mortar and add, by means of a burette sufficient water for it all to be soaked in the powder (no liquid can be seen outside the powdered mass). Repeat the process with a new quantity of powder and add the same number of ml of solution containing the desired quantity of the element to be determined. Dry under an infrared lamp and collect the dried mass with a small spatula. Loss of added material is minimized in this way. Intensive powdering is not necessary. Because of de-mixing with time (the added material sticks to the outside of each particle of the sample) such preparations are not tenable for long.

4.1.3. *Placing the loaded anode and the cathode in the electrode holders*

It has been mentioned in Section 1.1, that the anode of a d.c. arc reaches the highest temperature and as the method of analysis described in this book aims at a complete volatilization of all kinds of materials, the loaded anode is placed underneath and the cathode upside down above it.

In Section 1.4.3 (see (2)), maximum temperatures of the anode tip are given (3500–4000 °C). After arcing, the anode has become almost entirely red hot and for this reason, chrome iron clamps are used. They are pressed around the anode by means of a spring (see Fig. 15). Clamps and electrodes are put into contact with each other over 10 mm. In this way heat transport is fixed. If the clamps are water-cooled, heat transport is so large that it is entirely impossible to evaporate $Al(_2O_3)$.

The clamps have to be cleaned thoroughly, particularly the cathode holder, as vaporized material condenses on it (intermediate cleaning with a piece of wadding soaked in alcohol).

4.1.4. *Different kinds of graphite*

Graphite rods of British, American, German and Czechoslovakian origin show a specific weight (determined by weighing and measuring length and diameter) of 1·65. The German firm Ringsdorff also prepares rods of density 1·78–1·80. This material is not so reactive as the other specimen and it can be observed by heating rods in air for an hour at 900 °C. High density graphite rods lose 15 % less in weight than the others.

Decreased reactivity of high density graphite with metals like W and Ta is appreciable, spectra are stronger. Correction factors for this effect are mentioned in Section 4.2.2.

4.1.5. *Unexpected spluttering of the analysis material during arcing*

Spluttering can take place:

(a) at the moment when the arc begins to adhere to the sample. As particles are thrown out of the crater, a regular visual control of the arc (red glass window) is necessary. Faint red glowing particles can sometimes be observed at the beginning of arcing; they are of no significance as they originate from the graphite powder used.

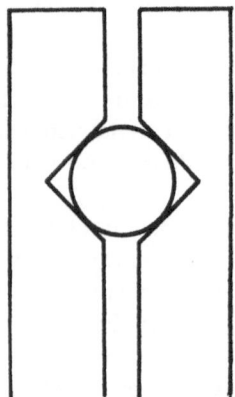

Fig. 15. Schematic drawing of electrodes (circle) placed in clamps. Electrodes and clamps touch each other over a distance of 10 mm.

(b) during arcing, when volatile compounds or elements have already been evaporated.

(c) at the end, when only less volatile material remains in the anode. For instance an aluminium alloy leaves some Al_2O_3 at the end of the exposure.

Usually such an analysis will be repeated. Spluttering can have taken place by chance. Otherwise precautions can be taken by preparing less coarse material or by covering the sample with a larger quantity of graphite powder (instead of 5 mg: 15 or 30 mg). But what can be done if no more material is available? In such cases it is of importance to know when spluttering has occurred.

The three cases a, b and c mentioned above are discussed in detail for the K-arc and the Q-arc as follows:

(a) (K-arc). Because of loss of the original sample, it is as if less than 5 mg has been weighed in. Therefore after complete analysis of the spectrum, the result of which delivers a sum of $x\%$ ($<100\%$), all concentrations calculated have to be multiplied by $100/x$. By experience it has been found many times that these results agree well with those obtained by measuring spectra where spluttering has not taken place.

(b) (K-arc). The spectrum obtained is of no value.

(c) (K-arc). Apply the E.C.F. (Section 3.3.1) for all elements except for the remaining refractory material.

(a) (Q-arc). All results are too low.

(b) (Q-arc). The spectrum obtained is of no value.

(c) (Q-arc). See c (K-arc).

Although in case of the Q-arc complete evaporation is also aimed at, visual observation of spluttering or of remaining material after the whole exposure is decisive.

4.1.6. *Unknown samples. General prescription*

K-method; (anode Fig. 13.1) mix and cover with 7 mg graphite powder; exposure time 7 minutes.

Q-method; (anode Fig. 13.5) exposure time 1 minute; if the evaporation is incomplete (visual observation) an exposure time of 2×1 or 3×1 minute should be chosen (spectra in successive steps of one minute; see remarks Table 20).

It may be that the results of the measurements indicate that no more precautions need to be taken, because I.C.F. and E.C.F. (*K*-method) agree with each other. If not, incomplete evaporation usually has taken place and data mentioned at the end of *K*- and *Q*-tables of each element (main component)—see Appendix— should be consulted. Furthermore, details given in Section 4.2.2 have to be borne in mind.

4.1.7. *Details concerning the shape and the loading of the anode and the exposure time chosen*

In most cases the kind of material that has to be analysed is known; if not, spectra taken according to Section 4.1.6 give sufficient information. It is even possible that no more taking of spectra is necessary. One can check this by looking through the comments at the end of *K*- and *Q*-tables of each element (see Appendix and also Section 4.2.2), as has been mentioned before.

Abbreviated data given at the end of each table. What do they mean? Let us take the metal Pt. K, 13.2; 5 Ni m; 7 C; 8, means: *K*-method; anode shape Fig. 13.2 (cathode shape always Fig. 13.3); load the anode with 5 mg Pt to be analysed (prepared in a finely divided form by cutting or filing, followed if necessary by cleaning with acid—see Section 4.1.2); the *K*-method used indicates that 5 mg Pt has to be taken. Now add 5 mg Ni powder and mix well with the sample; this is the meaning of: 5 Ni m. (If *m* has been left out (–5 Ni–), it means that 5 mg Ni powder has to be placed first in the anode and afterwards 5 mg of the sample on top of the nickel powder.) As has been pointed out in Section 4.1.1 it is necessary to cover the mixture of Pt sample and Ni powder with graphite powder (7 mg graphite powder: 7 C). This addition promotes smooth burning. In cases where one has to deal with more volatile substances to be analysed, 30 mg graphite powder is added (see Table 19); this large quantity prevents spluttering. The last figure (8) means that the exposure time amounts to eight minutes.

In the case of Pt, an indication is given at the end of the table that another method of burning of the arc can be applied also, namely: K, 13·4; ——; ——; 4, which means: fill the Harvey anode (Fig. 13.4) with 5 mg of the metal to be analysed; additions of Ni- and graphite powders have been omitted; the same cathode (Fig. 13.3) is applied and the exposure time is four minutes. *K*-factors in the Pt-table are given for this method of burning the arc. As one is mostly interested in impurities of the other platinum metals (Ru, Rh and Pd; Os and Ir) K-factors of these metals, determined by applying the Harvey-anode (Fig. 13.4), have been given there also.

The reader will observe that K-factors determined by applying the anode shapes (Fig. 13.2) are also given in the tables. This is because although the platinum metals with high boiling points can be evaporated more easily (and more completely) from anodes with a smaller diameter, K-factors of all the other metals have been determined by applying the anodes 13.1 or 13.2.

Up till now we have discussed the reason for the addition of graphite and nickel. However, more substances are added to the sample, these are Li_2CO_3 and $BaCO_3$. A substance like Li_2CO_3 is used for stabilizing the temperature of the arc (determination of the alkalis) and further for a decrease of the intensity of the cyanogen bands (these bands originate from the core of the arc and Li decreases the temperature of the core).

$BaCO_3$ well mixed with the sample establishes a smooth evaporation particularly when non-conducting oxides have to be analysed. Barium showing a low ionization potential, delivers free electrons easily and the arc quickly adheres on the sample so a more complete evaporation is obtained. The formation of Ba-salts like Ba-molybdate, Ba-niobate, Ba-tantalate, etc., is not excluded during heating of the anode. Later on, these salts dissociate completely into the elements at the arc gas temperature.

It can be pointed out again that Q-factors are applicable for both anode shapes 13.1 (13.2) and 13.4 (see Section 4.1.1).

An exposure time of 3×1 (see the K- and Q-tables) means that three separate spectra each of 1 minute exposure time have been taken (the arc burns continuously). These are known as time-resolved spectra. A lower background and a better orientation in the spectra so obtained are achieved in this way. Intensities of a spectral line to be measured, found in each spectrum have to be added up and their sum is a measure for the concentration of the element in question (application of K- or Q-factors and I.C.F. or E.C.F.).

Note. Quantities of 7, 15 or 30 mg graphite powder, 5 mg nickel powder, 10 mg Li_2CO_3 or $BaCO_3$ have to be weighed-in approximately. However, 5 or 10 mg of the sample has to be weighed carefully, for instance by means of a torsion balance checked by a standard weight of 5 or 10 mg. Mixing the added substances with the sample to be analysed can be carried out in the anode using a small spatula.

4.2. Explanation of the tables of K- and Q-factors of 63 elements (see Appendix) including important comments for calculating results of the analysis

This section has been divided into two parts: part 4.2.1 explaining terms, abbreviations, etc., mentioned in the tables of K- and Q-factors including values of these factors for the analysis of glasses (see p.88); part 4.2.2 which gives information with references to Sections 2.4, 2.5 and 2.6.

Wavelengths and values of transition levels have been taken from [14]. For the rare earths wavelength tables compiled in [24] have been consulted. Some values of excitation potentials have also been taken from [18] and [29].

K- and Q-factors of wavelengths greater than ~4000 Å have been found by applying photographic emulsions other than Kodak SA2 or SA3 (for instance Kodak 1N). Ilford plates or other plates equivalent to Kodak plates SA2 or 3 as to their H and D curves can also be applied.

4.2.1. At the head of the table of each element have been mentioned the following items:

(i) The symbol of the element and its name;

(ii) V_i its ionization potential in eV, which gives information as to the degree of ionization of the atom if present in the arc and also the origin of the radiation of the excited atom (core or mantle; see Fig. 5 and Section 1.6).

(iii) The ratio $Q/K(I)$ for spark lines and $Q/K(II)$ for spark lines (see summary at the end of Section 1.6 and the formulae 11 and 12 in Section 2.2). These formulae can also be used by approximation for control of Q-factors found experimentally and if only the latter have been determined, for a rough estimation of K-factors. Exceptions: in case of incomplete evaporation or (strong) self-absorption.

(iv) The boiling point (b.p.) of the element or a derivative of it (in °C).

(v) S_i the speed of evaporation of the element in question (its oxide or its carbide); see Eq. (2) and Fig. 4 (Section 1.4.3). In most cases the value of S_t is not to be taken into account for calculating analysis results; K-factors mostly include the S_t-value of the element in question. There are, however, exceptions. See Section 1.4.4, Section 2.6 and Section 4.2.2.

(vi) A: the atomic weight;

(vii) $O.f.$: conversion factor for calculating the concentration of the Oxide from the concentration of the element found;

(viii) L2–8000 (see [23]): the number of spectral lines between 2000 and 8000 Å in order to have an idea of the complexity of the spectrum in question beforehand. Examples: Li (39 lines), Na (175 lines), Ca (662 lines), Cu (913 lines), Rh (1327 lines), Ti (2136 lines), Ir (2577 lines), Mo (3902 lines), W (4327 lines), Fe (4757 lines) and Ce (5755 lines). Resolving power of present-day spectrographs is large enough to separate even the lines of Ce, but more complex spectra are obtained if more than one element is present as a main constituent.

(ix) The head of the second column of the table indicates the potential of high and low levels of excitation. Apart from the influences caused by excitation temperature, transition probabilities, etc. (see Section 2.1) K-factors will increase, as V_e (the high level of excitation) increases. Some control of the exact value of the K-factor of the element in question is possible in this way. If the value of the low level of the atom is zero or nearly zero, there is a fair chance that self-absorption (Section 2.3) can take place. However, if the atom population in the measuring volume is small as it is in the case of high boiling substances, self-absorption will not be observed; see platinum. For low boiling substances (zinc as an example or the alkalis) the effect of self-absorption is relatively large, as is indicated by a level of 0 eV. High levels of V_e (of arc lines) greater than V_i of the element in question indicate that one has to deal with two-electron transitions (e.g. see aluminium). Experience shows that such lines give less reproducible results.

(x) The fourth column, indicated in the head by 'Intensity' or 'Int.', has been added in order to indicate limits of intensity measurements (K-method). Low values are preferable because of the fact that these lines are narrow and do not show the effect of self-absorption. In cases of sparse spectra, extension of intensity measurements has appeared to be necessary in order to determine higher concentrations.

(xi) The fifth column contains Q-factors calculated from K-factors mentioned in the third column by applying values Q/K given in the head of the table (see further what has been said about $Q/K(I)$ and (II) at the beginning of this section).

(xii) The seventh column indicated by 'Remarks' contains a few indications concerning coincidence (further consultation of [19], [23] and [30]). Indications are as follows: coinc. Fe, which means: there is a coincidence of the line in question with an iron line; or coinc. $\geqslant 50\%$ Pb, which means that 50% Pb or less present in the sample does not influence the intensity measured.

(xiii) Other indications I-40 mean that Q-factors can be applied up to intensities of 40 (density $= 1\cdot1$). Furthermore, matrices in which the element in question has been determined, have been defined. If other matrices are analysed for the same element, consult Section 4.2.2. In case of self-absorption, tan α'' has been given (see Fig. 6a and b, Section 2.3).

(xiv) At the end of each element, table indications are given concerning loading of the anode, its shape and the exposure time as described in Section 4.1.7.

4.2.2. *Important considerations when calculating analysis results, and other details*

Because of matrix effects described in Sections 2.4, 2.5 and 2.6, correction factors are given below.

Ag   K (Ag 2721) $= 8\%$ if the light source slit is made twice as large, therefore twice $\frac{1}{2}$ mm. K (Ag 3280) and K (Ag 3382) have been determined in ZnS as a matrix ($S_t = 260$); thus for Pb as a matrix ($S_t = 180$) calculated, results $\times 180/260$ (Section 2.6).

       Stronger self-absorption in case of the application of K-factors causes a discrepancy between Q-factors calculated and experimental.

       K-factors, traces of Ag present in C or Se: results $\times 2\cdot5$ (Section 2.5).

       Q-factors, traces of Ag present in C or Se, results $\times 10$ or 4 respectively (Section 2.5).

       As to the evaporation of Ag (Section 1.3).

Al   Calculated concentrations (K-factors) of *impurities present in Al* have to be corrected as follows:

	Principal cause see
Si $\times 1\cdot2$	Section 2.4
Cu $\times 1\cdot4$	Section 2.4
Pb $\times 0\cdot25$	Section 2.6
Fe $\times 1\cdot1$	Section 2.4
Mn $\times 2/3$	Sections 2.4 and 2.6

Al_2O_3 present in $BaTiO_3$, results $\times 0\cdot75$ (Section 2.5).

Al present in magnet steel, results $\times 1\cdot6$ (spluttering, Section 2.6).

As   *K*- and *Q*- factors have been determined in graphite + impurities, *K*-factors of the lines 3032, 3119 and 2990 in $Mg_3(AsO_4)_2$.

K (As 2780) and (As 2860), results $\times 1 \cdot 5$ (Section 2.4 if As is present in glass).

Arsenic present in W or WO_3, Q (As 2860) = $0 \cdot 045 \%$ (Section 2.6, stable As-W compound).

Arsenic present in Sn or Pb-Sn, results $\times 1 \cdot 25$ (Section 2.6, spluttering).

Au   Section 1.3 for evaporation of gold.

B   If present in Si or Si-compounds (glasses), subtraction of neighbouring SiO bands is necessary. In order to determine relatively high concentrations a weakening sector of 1:30 has to be applied (rotating sector).

Q-factors of both *B*-lines, results $\times 1/3$ in matrix Si(C) (Section 2.6).

Ba   If Ba-metal has to be analysed, conversion into $BaCO_3$ is necessary (expose to the air).

K-factors have been determined by analysing carbonates or sulphates of Ba, Ca or Sr or, their mixtures.

Q-factors. No corrections are needed, except in those cases where extremely small amounts of material are analysed, *Q*-factors of spark lines $\times 0 \cdot 1$ (see Section 2.5).

Be   *K-method*. For higher concentrations, the application of a weakening sector of 1:30 is necessary. *K*-factors have been determined in Be-Cu alloys and Be containing zinc silicates.

Caution. Be-vapour is extremely poisonous. Control fume-hood by evaporating MoO_3.

Bi   If Bi is present as Bi_2Te_3 (volatile Te, Section 2.6), results have to be multiplied with $1 \cdot 5$; (*K*-factors of 3034, 2515 and 1798).

Ca   *K-factors of arc lines*. The influence of NaCl (in excess), Al_2O_3, $BaCO_3$ and $SrCO_3$ (all these elements have a V_i between 5 and 6 eV) cause a shifting of the atom/ion equilibrium $Ca \rightleftharpoons Ca^{++} + 2e$ to the left. *K*-factors have therefore in those cases to be multiplied with $0 \cdot 7$ (Section 2.5).

In calcium phosphates (V_i of P is high: 11 eV) *K*-factors have to be multiplied by 1.1.

No correction is needed for Ca present in GeO_2.

Q-factors of spark lines (Section 2.5). Traces of Ca present in C: results $\times 0 \cdot 05$; present in GeO_2: $\times 0 \cdot 1$.

Cd   No remarks.

Ce   *K*-factors (arc lines) have been determined in mixtures of CeO_2 and C, CuO or $CaCO_3$ respectively. In mixtures of CeO_2 and $BaCO_3$: *K*-factors $\times 0 \cdot 5$ (Section 2.5).

Co   No remarks.

Cr   Chrome steels as well as Cr_2O_3 have to be mixed thoroughly with C in order to prevent spluttering. Measurements of arc *and* spark lines have to be averaged (Fig. 5).

Cs   No remarks.

Cu   If present in brass: *K*-factors $\times 1 \cdot 2$ (S_i of zinc is high); if present in iron or steel: K (Cu 2824) and K (Cu 2961) $\times 0 \cdot 6$ (Section 2.6, S_i of Fe = 40, S_i of Cu = 65).

Sometimes results of measurements of the resonance lines and Cu 2824 do not agree. This is caused by diverging S_i and self-absorption (see under the head, 7th column Remarks). For very low Cu concentrations the *Q*-method is preferred; for somewhat higher results the *K*-method.

Dy   *Q-factors* have been determined in CaF_2 as matrix.

Er   *K-factors* have been determined in $ErSi_2$.

Eu   *Q-factors* have been determined in $PbCl_2$ as matrix.

Fe   *K-factors.* In case of iron oxide ($S_i = 60$) mixing thoroughly with C is necessary in order to stimulate reduction to Fe ($S_i = 40$). In many cases (for example materials sintered at high temperatures) results are too low, *inter alia* because of spluttering. Therefore results of relatively volatile elements (Cu, Sn, Pb) have to be multiplied by the E.C.F. (Section 3.3); less volatile element concentrations found have to be multiplied by the I.C.F. (Section 3.3). A substance like zinc ferrite cannot be analysed for its main components, because of the uncontrollable spluttering of volatile zinc.

     Q-factors. Traces of Fe present in P: results ×4 (carrier effect, Section 2.6).

Ga   *K-factors* have been determined in GeO_2 as matrix.

     For matrix = $In_{(2}O_3)$: *K*-factors ×0·4 (Section 2.4 and 2.5).

     For matrix = ZnO: *K*-factors ×2·5 (Section 2.6; S_i).

     For matrix = CdS: *K*-factors ×1·5 (Section 2.6; S_i).

     For matrix = Au: *K*-factors ×2 (Section 2.6; S_i).

Gd   *K- and Q-factors* have been determined in mixtures of CdS and Gd_2O_3 (0·04%–100%).

     Q-factors. Gd present in PbO ($S_i = 85$), results ×0·7 (S_i CdS = 160).

Ge   *K-factors* of 2651–3296 have been determined in a 5 mg sample mixed with 10 mg $CaCO_3$ and covered with 7 mg C; exposure time 7 minutes. *K*-factors of 2533–3067 have been determined as described in Section 4.1.7 (K; 13·1; 7 Cm; 7 C; 7). If Ge is present in In these *K*-factors have to be multiplied by 1·25 (Section 2.4).

     Q-factors of 2651–3269 are not significantly influenced by the absence or presence of $CaCO_3$.

     Speed of evaporation

     Ge-metal. S_i found = 20; (Fig. 4) boiling point = 3700 °C; there were lines in the X-ray diffraction pattern of the incompletely evaporated product which could not be ascribed to Ge, GeO_2 or graphite. In mass spectrometrical measurements [21], there are also indications for the existence of GeC.

     GeO_2. S_i found = 60 (Fig. 4); boiling point = 2700 °C, indicating that GeO_2 has been reduced in the anode to Ge (boiling point = 2700 °C).

     From the fact, experimentally found, that without the addition of $CaCO_3$ the *K*-factors (2651–3269) of Ge present in pure SiO_2 have to be multiplied by 3 (=60/20), the conclusion may be drawn that $CaCO_3$ favours the formation of Ge, whilst in pure SiO_2 the carbide has been formed.

Hf   No remarks.

Hg   *K- and Q-factors* have been determined in C + impurities and in HgS. HgS present in CdS (20% HgS, 80% CdS): *K*-factors × 0·5 (carrier effect, Section 2.6).

In   *K- and Q-factors* of 3256–2710 have been determined in matrices like Ni, GeO_2 and CdSe. In matrices like $CaCO_3$ and Ga_2O_3 (Section 2.5): results ×0·3.

     K-factors of 2713–3187 have been determined in In.

Ir   No remarks (Section 4.1.7).

K   *K- and Q-factors* have been determined by applying anode shape in Fig. 13.2 after covering with 10 mg Li_2CO_3. At normal arc temperature (no Li_2CO_3 added and at low concentrations of K): results ×3 (Section 2.5).

La   *K-factors* of 3245 II and 3303 II results × 1/3, if La is present in $BaTiO_3$ (carrier effect; Section 2.6).

Li   *K-factors* have been determined with an excess of K_2CO_3 (10 mg) added to a 5 mg sample. With less alkali present (for instance 10% as in the mineral lepidolite) results ×2 (Section 2.5). If no other alkali metal is present (Li-ferrites, Li-Cu alloys) *K*-factors ×2·2 (Section 2.5).

         Q-factors. If the other alkali metals are absent results × 2·5 (Section 2.5).

         Note. In all three cases shifting of the equilibrium $Li \leftrightarrows Li^+ + e$ takes place to the right, less Li-atoms remain (Section 2.5).

Mg   *Evaporation* as the oxide (reduction with C at 2200 °C); $S_i = 100$, as metal (boiling point of Mg 1100 °C); $S_i = 280$ (Fig. 4).

         K-factors. Mg present as metal in Fe or Ni; $S_i = 280$. Mg present in Al; $S_i = 100$ (evaporation via the oxide phase).

         Examples of *K*-factors:

		Main component		Main component
K 2852	·0006%	Glass, Al-silicate	·002%	Ni(O), Fe
2795	·0006	Glass, Al-silicate	·002	Ni(O), Fe
2802	·0015	Glass, Al-silicate	·0025	Ni(O), Fe
2779	·032	Glass, Al-silicate	·06	Ni(O), Fe

All the other *K*-factors have been determined in Al, glass, Al-silicates and MgO ($S_i = 100$).

         Note. The determination of small amounts of Mg(O) present in Ni leads to results depending upon the degree of oxidation of Mg present.

         Q-factors have been determined in oxidic materials.

Mn   *Spluttering.* Pure Mn metal as well as Mn present in Fe can show spluttering. Finely dividing the sample (extra small filings) and mixing it thoroughly with C are necessary.

         Excitation. If possible, use results of measurements of arc lines as well as of spark lines. Arc lines originate from the core as well as from the mantle (Fig. 5), spark lines are excited in the core.

         K-factors of 2738–2694 have been determined in MnO_2 and $MnCO_3$. *K*-factors of 3044 I, 3073 I and 3070 I have to be corrected as follows: Mn present in Al: ×2/3 (Section 2.5). Mn present in Zn-salts: ×3½ (Zn carrier, Section 2.6).

Mo   *General.* K- and Q-factors of Mo present in steel have been determined. In the oxide form, results have to be multiplied by 0·55, caused by the fact that MoO_3 is a volatile substance (Section 1.3, tungsten).

         K-factors. In addition to the *K*-factors mentioned in the table the following results have been found (matrix MoO_3; $BaCO_3$ added): K 2726·97 = ~34%; K 2885·73 = ~40%; K 2670·32 = ~48% and K 2640·28 = ~48%.

         In Na_2MoO_4 no corrections according to Section 2.4 have to be made (no simultaneous evaporation of the main constituents).

         K 2816 II in Ta ($S_i = 14$) as a matrix = 0·1% (Section 2.6).

         Q-factors. Q 3170 in W as a matrix = 0·0024%.

Na   *General. K-* as well as *Q*-factors of the lines 3302, 3303 and 2852 have been determined in matrices having $V_i > 7$ eV — *K*-factor of 2680 has been determined in a matrix of Na_2CO_3.

   K-factors of 3302, 3303 and 2852 have to be corrected according to Section 2.5, as follows: potassium oxide present in the sample for

$$\leqslant 1\%: \text{ results } \times 0\cdot 8$$
$$\sim 4\%: \text{ results } \times 0\cdot 6$$
$$\sim 8\%: \text{ results } \times 0\cdot 4$$
$$\sim 10\%: \text{ results } \times 0\cdot 3$$

   Q-factors of 3302 have to be multiplied by 0·2 if sodium is determined in a Li-matrix.

   Note. It is recommended **not** to measure high intensities (>8).

Nb   *K-factors* (low concentrations) have been determined in steel; they have to be multiplied by 0·6 if Nb is determined in a W-matrix.

   For oxide mixtures (for instance of Nb_2O_5, Ta_2O_5 and Fe_2O_3) mix well with 10 mg $BaCO_3$ (anode Fig. 13.2; exposure: 8 minutes). The following lines are recommended to measure:

$$2627\cdot 44 \text{ I } \quad 5\cdot 04–0\cdot 35 \text{ eV} \quad K = 4\tfrac{1}{2}\%$$
$$2649\cdot 51 \text{ I } \quad 4\cdot 90–0\cdot 27 \text{ eV} \quad K = 5\%$$
$$2782\cdot 36 \text{ I } \quad 5\cdot 59–1\cdot 15 \text{ eV} \quad K = 6\%$$

   Q-factors have been determined in graphite mixtures.

Nd   The boiling point of the oxide is high (>4200 °C). If the oxide is present in $AlPO_4$, its evaporation starts after two minutes burning of the *Q*-arc.

   Q-factors have been determined in a matrix of $CdWO_4$; in a matrix of LaF_3: *Q*-factors × 0·85.

Ni   *K-factors.* Ni present in Mo: K 3050 = 0·013%. Ni present in Al: K 3315, 2992, 3105 and 3145: ×5/3; (Section 2.4).

Os   No remarks (Section 4.1.7).

P   *K-factors* have to be multiplied by 4 in K_3PO_4 as matrix; by 2 in Ca- or Sr-phosphate (Section 2.4). If present in zinc phosphate results ×3/4 (S_i of *P* has been decreased, Section 2.6).

   Q-factors have been determined in Cu, SiC or GeO_2. The 2554 line is less coinciding with Fe-lines than the other three *P*-lines. There is an indication that in Fe the *Q*-factor 2554 amounts to 0·17% P (sputtering (Section 2.6)).

Pb   *Evaporation*, Section 1.4.4.

   K-factors 2833–2577 (determined in metals) have to be multiplied by 0·5 if the sample is oxidic. For higher concentrations (2628–2657) there in no difference, apparently because the oxides are always reduced to Pb.

   Pb can also show sputtering; Pb present in steel: *K*-factors × 5/4 (Section 2.6).

   In order to obtain reproducible results, it is necessary to have the sample, to be analysed, in a finely divided form (filings not drillings). Pb ($S_i = 180$) present in Al ($S_i = 100$), *K*-factors should have to be multiplied by 100/180 (Section 2.6, carrier effect), but because of gradual oxidation of Al to Al_2O_3 ($S_i = 25$) a factor of 0·3 has to be applied.

   Note. Intensity values (1–5) of the first four lines may not be exceeded.

Pd   If Pd is present in Ag (80% Ag, 20% Pd) it is carried into the arc with a higher speed (S_i Ag = 120; S_i Pd = 90) and *K*-factors (anode 13·2) have to be multiplied by 4/3 (confirmed experimentally for the lines 3142, 3009 and 3218) (Section 4.1.7).

Pr *Q-factors* have been determined in $AlPO_4$ as a matrix. One has to be very careful as all the lines mentioned are situated in cyanogen bands.

Pt No remarks (Section 4.1.7).

Rb No remarks.

Re *K-factors* of the lines 2999–2992 have been determined in a matrix of W by applying anode shape Fig. 13.4; those of the other lines by putting 5 mg Ni underneath (also anode shape Fig. 13.4); matrix Re.

Rh No remarks (Section 4.1.7).

Ru No remarks (Section 4.1.7).

Sb *Q-factor* of 2598 if Sb is present in Ge: 0.0014%.

Sc *Q-factors* have been determined in ZnS as a matrix.

Se *Q-factors* have been determined in CdSe (Q 2413 = 2%); in a mixture of ZnS and ZnSe (Q 2547 = 7%); in CdSe (Q 2547 = 10%) and in a mixture of ZnSe and Se (Q 2547 = 15%).

Si *K-factors*. Matrix Al: K 2532 = 3% (Section 2.4) or occlusion by Al_2O_3?

$\quad$ *Q-factors* of 2516–2514 have been determined in PbO as a matrix.

Sn *K-factors*. Matrix Fe (Sn probably present as the oxide); K 3175 and 2839: ×1·7 (sputtering, Section 2.6). Matrix Zn; K 2812–2787: ×1·1 (carrier Zn, Section 2.6).

Sr *K- and Q-factors* have been determined in carbonates or sulphates of the alkaline earths. Matrix Sr-phosphate: *K*-factors × 1·6 (Section 2.4).

Ta *K-factors*. No temperature corrections (Section 2.4) have to be made if $BaCO_3$ is added in case of oxidic samples. If graphite rods of high specific gravity (for instance Ringsdorff graphite rods RW III) and low reactivity with metals are applied, all *K*-factors (anode 13·2) have to be multiplied with 0·65 (Section 1.3 and 4.1.4).

Te *K- and Q-factors* have been determined in the following matrices: Te, Bi_2Te_3, ZnTe, CdTe, SnTe and C + GeO_2.

Th *K-factors* of the lines 2692–2641 have been determined in a matrix of W (containing some ThO_2): anode shape Fig. 13.1; 5 mg Ni underneath; exposure time 8 minutes.

$\quad$ *K*-factors of the lines 3014–2498 have been determined in a mixture of 5 mg ThO_2 and 10 mg $BaCO_3$; anode shape Fig. 13.2, exposure time 8 minutes.

Ti *K-factors* of the lines 3299 and 2661 have been determined in magnet steels; those of the lines 2974 and 2812 in TiO_2 and $BaTiO_3$ as matrices.

Tl *K-factors* have been determined in PbO as a matrix.

$\quad$ *Q-factors* originate from measurements in C as a matrix. Q 2767 = 0.0016% if Tl is present in PbO (small amounts of Tl carried away by PbO; S_i = 85); if present in In: 0·001 % (see Section 2.5).

U *K-factors* have been determined in U_3O_8 + $BaCO_3$, also in uranyl acetate + $CaCO_3$.

$\quad$ *Q-factors* have been determined in silicates; if present in PbO, results ×0·2 (Section 2.6, PbO carrier).

$\quad$ *Note*. The d.c. carbon arc is not suited for the determination of traces of uranium (for instance the limit of detection in X-ray fluorescence analysis of less than 0·008 % has been found).

V No remarks.

W *K-factors* have been determined in steels. If graphite rods of high specific gravity and low reactivity are used (compare Ta and Section 1.3 and 4.1.4) multiply results (anode 13.1) with 0·5.

In case of oxides results have to be multiplied with a factor <1 (see Section 1.3 and 4.1.4); this factor amounts to about 0·5.

Y   *K-factors* of:
3242–3195 have been determined in PbO + Y-nitrate
3179–2919 have been determined in PbO or GeO_2 + Y_2O_3
2919–2956 have been determined in Y_2O_3
Q-factors have been determined in PbO + Y-nitrate.

Zn   *K-factors.* The determination of zinc present in refractory materials (see under Fe; zinc ferrites) cannot be trusted because of spluttering (Section 2.6).

Q-factors. If Zn is present in an alkaline matrix (Na or K) *Q*-factors have to be multiplied by $2\frac{1}{2}$ (Section 2.4 and Fig. 7).

Zinc matrix. Because of its high value of S_i, impurities present in zinc are difficult to determine accurately. *Q*-factors of the elements present as impurities in zinc have to be multiplied by $\sim$2. (Accurate results can only be obtained wet-chemically; not by X-ray fluorescence, because of a gradual change of the metal surface during exposure to the air).

Note. Because of its low boiling point and the high anode tip temperature of the *Q*-arc, sideward losses cause a discrepancy between *Q*-factors of Zn 3075 found by calculation and by experimental means.

Zr   *K-factors* have been determined in oxidic samples. If the anode (Fig. 13.4) is applied, results have to be multiplied by 0·3.

4.3. Accuracy and reproducibility

Accuracy takes into account systematic error (bias) and reproducibility (precision). Systematic errors caused by temperature fluctuations are relatively small. If excitation takes place at 6100 K (Fig. 2), it concerns elements with a $V_i \geqslant 7$ eV. Elements showing a lower ionization potential, temporary decreases of temperature during the evaporation of these elements (Section 2.4) as well as shifting of atom/ion equilibria (Section 2.5) increase systematic errors. This increase of bias is also caused by the carrier effect (S_i; Section 2.6).

Because of the 'inconsistency' of the d.c. arc just mentioned, twice the c.o.v. amounts to 2–10% (rel.)—see also the end of Section 3.3.1—provided that analyses are carried out in duplicate. In some cases accuracy may be even poorer. Compare this result with spark analyses or those by X-ray fluorescence [11] where 2 × c.o.v. amounts to 0·5% (rel.), but in these cases one has to deal with programmed analyses.

The advantage of the method of analysis described in this book is as follows: it gives a general view of all constituents of the sample with a reasonable accuracy according to a general method of handling for all kinds of material.

The accuracy is much less if the Q-arc is applied for the determination of very low concentrations; results can be trusted roughly within a factor of 2; in many cases, however, accuracy is better.

For convenience of the reader a simple method of determining the standard deviation $\sigma (= \pm \sqrt{\Sigma d^2/(n-1)}$; $d =$ deviation from the arithmetic mean and $n =$ number of measurements) is as follows: after having determined the range w, (equal to the difference between the highest and lowest values) of a number of n determinations, apply the following equation: $\sigma = A_n \times w$, where $A_n = 0.89$ $(n = 2)$; 0.59 $(n = 3)$; 0.49 $(n = 4)$; 0.43 $(n = 5)$; 0.40 $(n = 6)$; 0.37 $(n = 7)$; 0.35 $(n = 8)$; 0.34 $(n = 9)$ and 0.32 $(n = 10)$.

The coefficient of variation (c.o.v.) is equal to

$$\{100\ \sigma/(\text{arithmetic mean})\}\ \%$$

Twice the c.o.v. is a measure for the 95% confidence level, apart from systematic errors.

4.4. Some conversion factors for calculating the concentration of oxides and salts from the elemental percentages determined spectrochemically (A = atomic weight)

A	Oxides/salts	Conversion factors
107·9	Ag–Ag$_2$O	1·073
27·0	Al–Al$_2$O$_3$	1·886
75·0	As–As$_2$O$_3$	1·32
	As–As$_2$O$_5$	1·533
197	Au–Au$_2$O$_3$	1·12
10·8	B–B$_2$O$_3$	3·22
137·4	Ba–BaO	1·117
	Ba–BaS	1·23
	Ba–BaSO$_4$	1·69
	Ba–BaCO$_3$	1·435
9·0	Be–BeO	2·77
209	Bi–Bi$_2$O$_3$	1·115
40·1	Ca–CaO	1·4
	Ca–CaCO$_3$	2·5
	Ca–CaF$_2$	1·95
	Ca–CaSO$_4$	3·4
112·4	Cd–CdO	1·14
	Cd–CdS	1·28
	Cd–CdSe	1·70
140	Ce–CeO$_2$	1·23

A	Oxides/salts	Conversion factors
58·9	Co–CoO	1·27
	Co–Co$_2$O$_3$	1·41
52·0	Cr–Cr$_2$O$_3$	1·46
	Cr–CrO$_3$	1·92
132·9	Cs–Cs$_2$O	1·06
63·6	Cu–CuO	1·25
	Cu–Cu$_2$O	1·125
55·9	Fe–FeO	1·287
	Fe–Fe$_2$O$_3$	1·43
69·7	Ga–Ga$_2$O$_3$	1·34
72·6	Ge–GeO$_2$	1·44
178·6	Hf–HfO$_2$	1·18
200·6	Hg–HgO	1·078
	Hg–HgS	1·16
	Hg–HgSe	1·398
114·8	In–In$_2$O$_3$	1·21
39·1	K–K$_2$O	1·205
	K–KCl	1·91
138·9	La–La$_2$O$_3$	1·17
6·9	Li–Li$_2$O	2·15
	Li–Li$_2$CO$_3$	5·32
24·3	Mg–MgO	1·655
	Mg–MgCO$_3$	3·47
54·9	Mn–MnO	1·29
	Mn–MnO$_2$	1·58
	Mn–MnCO$_3$	2·09
96·0	Mo–MoO$_3$	1·5
23·0	Na–Na$_2$O	1·345
	Na–Na$_2$CO$_3$	2·3
	Na–NaCl	2·54
	Na–Na$_2$SO$_4$	3·09
58·7	Ni–NiO	1·27
31·0	P–P$_2$O$_5$	2·29
	P–PO$_4$	3·065
207·2	Pb–PbO	1·078
	Pb–PbO$_2$	1·155
	Pb–Pb$_3$O$_4$	1·1
195·2	Pt–PtO$_2$	1·163
121·8	Sb–Sb$_2$O$_3$	1·195
	Sb–Sb$_2$O$_5$	1·328
	Sb–Sb$_2$S$_3$	1·39
28·1	Si–SiO$_2$	2·135
	Si–SiC	1·43
118·7	Sn–SnO	1·134
	Sn–SnO$_2$	1·269

A	Oxides/salts	Conversion factors
87·6	Sr–SrO	1·181
	Sr–SrCO$_3$	1·68
	Sr–SrSO$_4$	2·095
180·9	Ta–Ta$_2$O$_5$	1·22
232	Th–ThO$_2$	1·138
47·9	Ti–TiO$_2$	1·665
51·0	V–V$_2$O$_5$	1·783
183·9	W–WO$_3$	1·259
88·9	Y–Y$_2$O$_3$	1·27
65·4	Zn–ZnO	1·243
	Zn–ZnS	1·49
	Zn–ZnSe	2·21
91·2	Zr–ZrO$_2$	1·352

4.5. Determination of blanks and of 'carbon impurities' present in 'pure' samples

4.5.1. When an arc burns between unloaded electrodes the evaporation of the 'carbon impurities' (Si, Mg, Fe, Al, Cu, (Ag) and B) takes place in such a way that a relatively large fraction does not reach the zone of excitation but is lost sideways in the stream of air ascending along the electrodes. This means that the investigation of spectroscopic graphite must not be carried out by having an arc burn between two clean electrodes. The material placed in the anode has a collecting action during evaporation and the effect of collection can be defined as the ratio of the intensities of spectral lines found with and without a material not containing the element in question. In practice extremely pure GeO$_2$ can be used. It is understandable that the time in which the material is evaporating will determine the collecting action. This effect, described for the first time in [33], and at greater length later in [31], is further described in detail in Section 4.5.2.

For the determination of graphite *blanks*, Q-analyses (Q; 13·5; 10 GeO$_2$; ——; 1) are carried out in six-fold; results of Si-determinations of a number of graphite rods will be given as an example. The mean concentration of the six determinations appears to be 0·00088 %;* w has been found to be 0·00084 (see Section 4.3); as

---

* It is recommended to express results in two or three decimals and to round off the final result.

A_n for six determinations is equal to 0·40, σ of a single determination amounts to $0\cdot40 \times 0\cdot00084 = 0\cdot000336$ and σ of the mean of six determinations $0\cdot000336/\sqrt{6} = 0\cdot000137$. We need this value for the determination of the detection limit of Si present in an unknown sample (Section 4.5.2 (i) and (ii)).

If we had not applied the addition of GeO_2 to the anode we would have found 0·00022% Si (collecting action of GeO_2 during a one-minute exposure amounts to 4) instead of 0·00088% and we should have subtracted the value of 0·00022% from the Si-content found in the unknown sample instead of 0·00088%. Let us suppose that the Si-content of the analysis sample had been found to be 0·0011%, the final result $0\cdot0011–0\cdot00022 = 0\cdot0009\%$ should have been sent to the customer instead of $0\cdot0011–0\cdot00088 = 0\cdot0002\%$.

The collecting action of GeO_2 during a one-minute exposure amounts to 4 and this effect has been shown to be fairly constant during the time of evaporation of this substance in the Q-arc: 42 seconds. During the remaining time of 18 seconds no collection takes place. How large is the real collecting action (p) during evaporation?*

Let us suppose that the contribution to the intensity of a spectral line amounts to i per second without the presence of a collector. We know that during a one-minute exposure the collecting factor of GeO_2 amounts to 4, which means that in one minute the intensity will be $4 \times 60 \times i$. During 42 seconds of effective evaporation of GeO_2 the contribution to the final intensity is $42 \times p \times i$ and during 18 seconds burning without collector $18 \times i$. Therefore:

$$42 \times p \times i + 18i = 4 \times 60i \quad \text{or} \quad p = \frac{111}{21} = \sim 5$$

Remarks

(1) The evaporation of GeO_2 is fairly constant; m.p. of GeO_2 and Ge are 1115 and 960 °C resp.; b.p. of GeO_2, Ge and GeC are 2700, 2700 and 3700 °C resp.; see Section 4.2.2 under Ge, and also [34]. The range of temperatures during which Ge atoms are entering the arc, is fairly large and their collecting action covers a large temperature range. For extremely high boiling substances it is advised to apply other substances, for instance SiC: b.p. $\geqslant$3800 °C.

* Here we have made use of a valuable comment of Drs. Millet and Kealy to whom we owe our sincere thanks.

(2) It is justified to assume that the real collecting action of all kinds of material during their evaporation interval is constant. For instance arsenic, evaporating during about 5 seconds through the Q-arc, shows a collecting factor of about 1·25. An analogical calculation as given above, gives a value of about 4 for p, derived from: $5 \times p \times i + 55i = 1·25 \times 60i$. The value of $p = 111/21$ is preferable because a measurement of 5 seconds (As) is less accurate than a measurement of 42 seconds (GeO$_2$).

4.5.2. The determination of impurities present in a pure sample is easy to be carried out as long as no 'carbon impurities' have to be taken into consideration. For 'carbon impurities' the following considerations hold.

Example: Once again analyses (in six-fold; three exposures on one plate and three on another!) of the unknown sample are carried out; its mean Si-content appears to be 0·000345%. It evaporates during t seconds. The contribution to the final intensity of a 'carbon element' present in the unknown substance is therefore

$$t \times \frac{111}{21} \times i + (60 - t)i$$

And as we wish to know the collecting action of this substance derived from experiments with GeO$_2$ the following ratio holds:

$$\frac{t \times \dfrac{111}{21} \times i + (60 - t)i}{42 \times \dfrac{111}{21} \times i + 18i} = \frac{t + 14}{56} \qquad (21)$$

and it corrects for the real blank to be subtracted. It has to be applied to the mean of six determinations, 0·00088% Si found in the carbon blanks +GeO$_2$. If the unknown sample shows for instance an evaporation time $t = 28$ seconds, $\dfrac{t + 14}{56}$ amounts to $\frac{3}{4}$ and the real blank to be subtracted is $\frac{3}{4} \times 0·00088 = 0·00066\%$ Si. Subtraction of this value from 0·000345% (the mean of six determinations of the sample) delivers a negative result. And we may conclude that Si is absent in the sample.

A summary is given in (i):

(i) *Determination of presence or absence of 'carbon impurities'* (Si as an example)

Mean concentration of Si found in a sample: 0·000345%.
Mean concentration of Si present in blank (with GeO_2): 0·00088%.
Correction for the same time $t = 28$

$$\frac{t + 14}{56} \times 0·00088 = \tfrac{3}{4} \times 0·00088 = 0·00066\%$$

Conclusion: Si is absent.

As a negative result (0·000345%–0·00066%) is nonsense one wishes to know the real limit of detection of Si present/absent in that particular sample.

In Section 4.5.1 it has been shown that σ of the mean of six blank determinations (with GeO_2) amounted to 0·000137. If we include the factor

$$\frac{t + 14}{56} = \tfrac{3}{4}$$

for $t = 28$ s of the sample in question σ blank amounts to $\tfrac{3}{4} \times$ 0·000137 = 0·000103. (It is assumed that w, σ (single) and σ (6-fold) will be diminished in the same sense.)

In the same way σ' of the mean Si-content of the sample can be calculated from $w' = 0·00022$, σ' (single) = 0·40 × 0·00022 = 0·000088 and σ' (6-fold) = $0·000088/\sqrt{6}$ = 0·000036.

The limit of detection of Si present in the sample to be analysed amounts to (compare also Eq. (18)) 2·2* times the sum of both σ's (quadratically added):

$$2·2\sqrt{(\sigma - \text{6-fold})^2 + (\sigma' - \text{6-fold})^2} =$$

$$2·2\sqrt{(0·000103)^2 + (0·000036)^2} = 0·0002^4\% \text{ Si}$$

This means that less than 0·0002⁴% Si can be present. In (ii) a summary is given.

---

* 95% confidence level and 10 degrees of freedom, cf. [35]; (12 measurements carried out in 2 series: 10 degrees of freedom).

(ii) *Determination of the detection limit of 'carbon impurities' present in a sample* (Si as an example)

GeO$_2$ blank

$w = 0.00084$ (see Section 4.5.1).

σ single $= 0.40 \times 0.00084 = 0.000336$

σ 6-fold $= 0.00036/\sqrt{6} = 0.000137$

σ 6-fold $\times \dfrac{t + 14}{56} = 0.000137 \times \frac{3}{4} = 0.000103$

Sample

$w' = 0.00022$ (see Section 4.5.2).

σ' single $= 0.40 \times 0.00022 = 0.000088$

σ' 6-fold $= 0.000088/\sqrt{6} = 0.000036$

limit of detection $= 2.2\sqrt{(0.000103)^2 + (0.000036)^2} = 0.0002^4\,\%$ Si.

It is interesting to compare results found with and without taking into account the collecting effect. (iii) Shows the results given in ppm.

(iii) *Comparision of results wrongly obtained by subtracting blank signal from sample signal and obtained by taking into account the collecting effect of the sample in question*

	Si	Mg	Fe	Al	Cu	
Sample (mean of 6 measurements)	3·5	1·5	1·6	0·6	0·12	ppm
Blank (mean of 6 measurements)	2·2	0·4	1·0	0·2	0·03	ppm
Concentration found by subtraction only	1·3	1·1	0·6	0·4	0·09	ppm
Real concentrations found after the method described above; all elements have been proved to be *absent*:	<2·4	<0·7	<0·8	<0·2	<0·04	ppm

It is interesting to realize that a high value of the limit of detection will be found if w (blank) is large, which can be caused by inhomogeneous distribution of the 'carbon impurities' in the graphite rods (or electrodes). If the sample is not homogeneous the same trend will be found. In order to obtain the best (the lowest) results homogeneity of both graphite and sample is necessary.

A summary of the content of this section is given in (iv) which can also be used in the practice of routine analysis.

(iv) *The determination of 'carbon impurities' (in ppm) present/ absent in a sample and their detection limits* (*Q*-method)

Plate numbers . . ., . . .; sample . . .; analysis number . . .; time of evaporation $t = . . .$ (*t* of the sample in question 28″); graphite rods, firm . . .; lot number . . .

		Si	Mg	Cu	Fe, Al, B, (Ag)
Concentration found in sample (Eq. (10)); E.C.F. applied	1	3·43	1·36	0·176 —	
	2	4·58 —	2·42 —	0·176	
	3	3·17	1·37	0·390 —	
	4	2·77	0·98 —	0·196	
	5	2·38 —	0·98	0·292	
	6	4·37	1·95	0·244	
Sample mean concentration		3·45 ←	1·51 ←	0·25 ←	
w′ (see Section 4.3)		2·20	1·44	0·21	
σ′ single = w′ × 0·40 (see Section 4.3)		0·88	0·58	0·08	
σ′ 6-fold = σ′ single/√6		0·36	0·24	0·03	
GeO₂ blank mean concentration		8·80	1·36	0·11	
The same ×(28 + 14)/56; Eq. (21)		6·60 ←	1·02 ←	0·08 ←	
σ 6-fold determined in the same way as for the sample		1·37	0·30	0·005	
σ 6-fold = the same ×(28 + 14)/56		1·03	0·23	0·004	
Mean concentration sample minus mean concentration GeO₂ blank corrected for *t*		negative ←	0·49 ←	0·17 ←	
Limit of detection: $2·2\sqrt{(\sigma'\text{ 6-fold})^2 + (\sigma\text{ 6-fold})^2}$		2·4	0·73	0·07	
Absent/present		absent	absent	present	
Final result		absent (<2½)	absent (<0·7)	0·17	

4.5.3. Sections 3.5 and 4.5 have been intended *inter alia* to fill the gap between spectrochemical and mass spectrometrical analyses of traces of elements.

List of abbreviations and definitions

A

A	atomic weight
α	degree of ionization
a	transition probability of a spectral line
α' and α''	slope of a log I/log C working curve

B

b.p.	boiling point
B	partition function; the sum of statistical weight factors for all energy levels except the one considered

C

c	velocity of light
C	concentration of an element (in weight %) present in a sample

carbon impurities; traces of elements present in graphite electrodes (Si, Mg, Fe, Al, Cu, sometimes Ag and B)

coincidence $>x\%$ of element y: the spectral line in question interferes only with a line of element y if this is present for more than $x\%$.

collecting action: the influence of the presence of evaporating substances causing minimal sideward losses of carbon impurities and therefore an increase of their line intensities

c.o.v.	coefficient of variation

D

d	deviation of one determination from the arithmetic mean of a number of elements
d.c.	direct current
D_b	density of the background near a spectral line
D_l	density of a spectral line
D_{l+b}	density of 'line plus background'
D_{ox}	degree of dissociation of a metal oxide

E

E	dissociation energy of metal oxide
E.C.F.	external correction factor
eV	electron volt

exposure time 3×1 minute: exposure in three separate steps of one minute each (the arc burns continuously)

G

g	statistical weight factor of an energy level; $(g = 2J + 1)$
g	gain factor; ratio of limits of detection applying s.p.s.- and Q-method of analysis

H

h Planck's constant

H and D curve: calibration curve of a photographic emulsion

I

i I per second

I intensity (properly speaking: energy) of a spectral line originating
                   from N_e atoms

I.C.F.   internal correction factor

I_b intensity of the background of (near) a spectral line

I_l intensity of a spectral line

I_{l+b} intensity of 'line plus background'

J

J internal quantum number

K

k Boltzmann's constant

K-factor: a proportionality factor equal to C, if I is unity (application of the
           K-method of analysis)

K-method: concerns the evaporation of 5 mg of a sample to completion (arc
            current 10 Amps)

K' ionization constant

K_T dissociation constant

K 13·1; 7 Cm; 7 C; 7 means:
           K,   method of analysis
           13·1,   anode Fig. 13.1
           7 Cm, $m$ix the sample with 7 mg graphite (C) powder in the anode
           7 C,   cover with 7 mg graphite powder (C)
           7,   exposure time 7 minutes

L

L 2–8000   the number of spectral lines of an element between 2000 and 8000 Å

λ wavelength of the emitted light

M

m.p.   melting point

N

n number of measurements

N_e number of excited atoms

N_0 number of atoms in the ground state

ν frequency of the emitted light ($c/\nu = \lambda$)

O

$O.f.$ conversion factor for calculating the concentration of the oxide from
             the concentration of the element found

P

p ratio of the signal of carbon impurities with and without the addition
             of evaporating substances other than carbon impurities

p_{rel} relative sensitivity of a photographic emulsion dependent on the
             wavelength

Q

Q-factor: a proportionality factor equal to C, if I is unity (application of the Q-method of analysis)

Q-method: concerns the evaporation of 10 mg of a sample (arc current 10 Amps)

Q 13·1; 7 Cm; 7 C; 3 × 1 means:

> Q method of analysis
> 13·1   anode Fig. 13.1
> 7 Cm   mix the sample with 7 mg graphite (C) powder in the anode
> 7 C   cover with 7 mg graphite powder (C)
> 3 × 1   exposure time 3 × 1 minute

S

S_t speed of evaporation

s.p.d. scale   standard paper density scale

s.p.s. method Q-method of superimposing spectra obtained by projecting the light source on the slit of the spectrograph (sometimes a 25 in lieu of a 10 mg sample evaporated)

σ standard deviation of a quantitative determination

σ' standard deviation of a quantitative determination of 'carbon impurities' present in a sample

σ_b standard deviation of the intensity of the background

σ_{l+b} the same for 'line plus background'

T

t time of evaporation and exposure

T temperature of excitation or ionization (K)

T_{bp} boiling point (°C)

T_{bpK} boiling point (K)

V

V_e excitation potential (eV)

V_i ionization potential (eV)

V_m ratio of measuring volumes in mantle and core

W

w the difference between the highest and the lowest values of a series of measurements (p. 75)

CONVERSION TABLE

Unit	SI Equivalent
Angstrom, Å	10^{-10} m
Boltzmann constant, k	$1·380 \times 10^{-23}$ JK^{-1}
Electron volt, eV	$1·602 \times 10^{-9}$
Centimetre, cm	10^{-2} m
Planck constant, h	$6·625 \times 10^{-34}$ Js
torr	$133·322$ Nm^{-2}

Bibliography

This list of papers is confined to those referred to in the text, and readers wishing to bring their knowledge of the subject up to date are recommended to scan the annual reviews of publications published in *Analytical Chemistry* and in England as *Spectrochemical Abstracts*.

1. Nickel, H., *Spectrochim. Acta*, 23B, 323 (1968); C.S.I. XIV, Debrecen 1967, pp. 467–8.
2. Rautschke, R., *Spectrochim. Acta*, 23B, 55 (1967).
3. Dikhoff, J. A. M., 1956 C.S.I. VI, Amsterdam, pp. 162–7. Pergamon Press 1957.
4. Addink, N. W. H., *ibid.*, pp. 168–78.
5. Addink, N. W. H., C.S.I. XIV, Debrecen 1967, pp. 201–8.
6. Addink, N. W. H., C.S.I. IV, Münster 1953. Proceedings have not been published.
7. Addink, N. W. H., C.S.I. V, Gmunden 1954, *Mikrochimica Acta Heft* 2–5, 1955, pp. 703–8.
8. Scott, V. D., *Nature* (London), 186, 466 (1960).
9. Addink, N. W. H., *Journal of the Iron and Steel Institute*, 194, 199 (1960).
10. Smit, J. A., *Thesis Utrecht*, 1950.
11. Addink, N. W. H., *Trace Characterization, Chemical and Physical* (1967), *N.B.S. monograph*. 100, pp. 121–48.
12. Boumans, P. W. J. M., *Thesis Amsterdam* 1961.
13. Addink, N. W. H., *Spectrochim. Acta* (1959), pp. 349–59.
14. Sitterly-Moore, Ch., *A multiplet table of astrophysical interest* (1945 edition); *An ultraviolet multiplet table*, circular 488 (Sections 1–5) and Technical Note No. 36 of the National Bureau of Standards, Washington D.C.
15. Corliss, C. H. and Bozman, W. R. (1962), *N.B.S.—monograph No.* 53.
16. van Hengstum, J. P. A. and Smit, J. A., *Physica* 22, 111 (1956).
17. Addink, N. W. H., *Spectrochim. Acta*, 9, 159 (1957).
18. Saidel, A. N., Prokofiev, V. K. and Raiski, S. M., *Tables of spectral lines*, VEB Verlag Technik Berlin, 1955 (p. xxxvii).

19. Ahrens, L. H. and Taylor, S. R., *Spectrochemical Analysis.* Pergamon Press, 1961, 2nd ed. (p. 27).
20. Saha, M. N., *Phil. Mag.*, **40**, 472 (1920.)
21. Köhl, G., Z.f. *Naturforschung* 9A, 913 (1954).
22. Scribner, B. F. and Mullin, H. R., *J. Res. Nat. Bur. Std.*, **37**, 379 (1946).
23. Harrison, G. R., M.I.T. *Wavelength tables.* John Wiley and Sons, New York 1939.
24. Norris, J. A., *National Bureau of Standards*, Institute for Applied Technology. Distr. by Clearinghouse. ORNL 2774, sec. I and II.
25. Addink, N. W. H., *Spectrochim. Acta*, **4**, 36 (1950).
26. Dikhoff, J. A. M. and Addink, N. W. H., *Mikrochimica Acta*, *Heft*, **2–3**, 257 (1955).
27. Eichhoff, H.-J. and Addink, N. W. H., C.S.I. VIII, Lucerne, 1959 *Proceedings. Verlag Sauerlander*, Aarau (Switzerland) 1960, pp. 89–92.
28. Addink, N. W. H., *Chem. Weekblad*, **56**, 622 (1960). In Dutch.
29. Grotrian, W., *Graphische Darstellung der Spektren von Atomen und Ionen mit ein, zwei und drei Valenz-elektronen* II, Springer, Berlin 1928.
30. Kroonen, J. and Vader, D., *Line interference in emission spectrographic analysis*, Elsevier Publishing Co. 1963.
31. Addink, N. W. H. and Witmer, A. W., C.S.I. IX, Lyon 1961, *Proceedings GAMS*, Muray-print. Paris 1962, pp. 340–54.
32. Kaiser, H. and Specker, H., *Z. Anal. Chem.*, **149**, 46 (1956).
33. Webb, D. A., *Nature* (London), **139**, 258 (1937).
34. Strock, L. W., *Appl. Spectroscopy*, **7**, (May 1953), No. 2.
35. Davies, O. L., *Statistical methods in research and production*, Oliver and Boyd, London, Edinburgh 1958, p. 366.
36. Pascal, P., *(Nouveau) Traité de chimie minérale*, Masson et Cie, Paris 1956–64.
37. Hodgman, Ch. D., *Handbook of Chemistry and Physics.* Published by Chemical Rubber Publishing Co.
38. Honig, R. E., *RCA Review*, Dec. 1962, Vol. 23, No. 4, pp. 567–86.
39. Van Stekelenburg, L. H. M., *Physica*, **12**, 289 (1946).
40. Schroll, E., C.S.I. XIV, Debrecen 1967, pp. 397–434.

Appendix

Glasses

1. *Determination of*: Al, As, Ba, Ca, Cd, Fe, Li, Mg, Mn, Na, Pb, Si, Ti and Zn.

K-method: 13·1; 7 Cm; 7 C; 7 (8). Weakening factor 1:6 (exposure 8 minutes if Al_2O_3 is present for more than 5%).

Element	Concentration range of the oxide	λ (Å)	K-factor (%) (unweakened line)
Al	1–5%	3082	0·01
		2567	0·14
		2575	0·14
		2660	0·2
		2652	0·35

Note 1. If about 30% PbO is present in the sample: results ×1·56 (S_t PbO = 85; S_t Al_2O_3 = 25; Section 2.6).

Note 2. If about 10% earthalkali oxides is present: results ×0·75 (Section 2.4).

Element	Concentration range	λ (Å)	K-factor
Ba	1–10%	3071	1·0
Ca	6%	3179 II	0·1
		3158 II	0·17
		3006	1·1
		2997	2·9
		3009	3·0
Fe	0·05–0·2%	3020	0·009
		2599 II	0·02
		2739 II	0·1
		2628 II	0·2
		2735	0·21
Li	0·5%	3232	0·1
		2741	0·6
Mg	3·5%	2852	0·0008
		2802 II	0·0015
		2779	0·03
		2781	0·1
		3091	0·5
		2941	1·7
Mn	0·3%	2798	0·0015 ⎱ (see
		2801	0·002 ⎰ Section 2.5)
		2933 II	0·04

Element	Concentration range of the oxide	λ (Å)	K-factor (%) (unweakened line)
Na	16%	3302	
	≤1% K_2O present:		0·27 ⎫
	~4% K_2O present:		0·2 ⎬ (see Section 2.5)
	10% K_2O present:		0·13 ⎭
		3303	
	1–10% K_2O present:		~0·4
		2852	
	1–4% K_2O present:		3
Pb	0–30%	2833	0·02
		2628	5
		3220	13
Si	60–75%	2532	2·5
		2568	13
		2970	16
Ti		3241 II	0·02

Note. Elements with a high value of V_i, such as Zn, Cd and As, present in low concentrations; results $\times 1\cdot5$ (see Section 2.4).

2. *Determination of*: boron (1 mg sample).
K-method: 13·1; 1 Ni; 7 C; 7. Weakening factor 1:30.
See: B(K- and Q-tables).
Results have to be multiplied with 5 (1 mg instead of 5 mg sample), further with 30 (if measurements have been carried out in the weakened part of the spectrum) and with O.f. (conversion factor $B \rightarrow B_2O_3$) and E.C.F. Because of the impossibility of summing up to 100%, the B-determination in glass is less accurate.

3. *Determination of*: potassium.
K-method: 13·1; 5 mg sample covered with 10 mg Li_2CO_3; weakening factor 1:6; exposure time 75 seconds.

Element	Concentration range of the oxide	λ (Å)	K-factor (%)
K	8%	4044	0·3
		4047	0·6

Note. Although Li_2CO_3 is added cyanogen bands cause a relatively heavy background and decrease the accuracy of measurements. For this reason comparison with standard samples is advised.

Index